THE SCHOOL MATHEMATICS PROJECT

CALCULUS AND ELEMENTARY FUNCTIONS 1

BY

R. M. N. MONTGOMERY

formerly of Winchester College

AND

T. A. JONES

Homerton College

CAMBRIDGE

AT THE UNIVERSITY PRESS

1970

Published by the Syndics of the Cambridge University Press
Bentley House, 200 Euston Road, London N.W.1
American Branch: 32 East 57th Street, New York, N.Y.10022

Library of Congress Catalogue Card Number: 74–108107

Standard Book Number: 521 07712 5

Printed in Great Britain
at the University Printing House, Cambridge
(Brooke Crutchley, University Printer)

CONTENTS

PREFACE

This book is intended as an introduction to the Calculus, and to those elementary functions with which the Calculus usually deals. This, for beginners, has often been a difficult subject; and we hope that where our approach is unconventional it may make clearer than usual the new ideas that can be perplexing, and reduce by contrast the number of new notations and new drills.

We have therefore throughout this volume used only the functional notation, $f(x)$, with $f'(x)$ for its derivative; and we have concentrated on what is in our opinion the essential drill of elementary calculus: that of sketching a function's graph. Moreover we have avoided the usual use of limits; but have preferred to calculate derivatives by the process of linear approximation (see Chapter 3) and the formula:

$$f(x+\epsilon) = f(x)+f'(x).\epsilon+(\text{terms small compared with } \epsilon).$$

(Thus, for instance, the fact that $f(x) = x^3$ implies

$$f(x+\epsilon) = (x+\epsilon)^3 = x^3+3x^2\epsilon+\text{terms in } \epsilon^2 \text{ and } \epsilon^3,$$

reveals that the derivative of x^3 is $3x^2$.)

Conceptually, this is the hardest part of the book; and its difficulty has not been disguised. We have adopted it because we believe that, once mastered, it is a more serviceable tool than taking limits for many of the elementary procedures of calculus, and that it throws more light on the ideas involved.

This notation and method, then, have been used to study the behaviour of algebraic, exponential and trigonometrical functions and those directly dependent on them. We have also shown how they can be applied to simple mathematical problems (such as solving equations); and made a start towards applying them to problems which arise in the physical world. The 'physical problems' are, naturally, of a highly simplified kind. Some discussion of 'mathematical modelling' is given in Chapter 4, Section 2.2; here we need only emphasize that we regard this 'simplification' as an essential tool of applied mathematics, and not as a facile device for producing 'problems' that can be worked into an elementary course.

This volume stops short of summation, differential notation, and all but the very simplest differential equations. It includes a short chapter on polynomial approximations.

R. M. N. M.
T. A. J.

January 1970

The [illegible] some discussion of [illegible] notation, and [illegible] the [illegible] polynomial approximations.

[illegible]

January 19[illegible]

1

FUNCTIONS

1. FUNCTIONS

The notion of a function is fundamental to mathematics. This book is concerned with certain simple types of mathematical function. There are many other types, but they do not concern us here. We shall therefore, in this first chapter, talk simply of those characteristics of functions which will be important through the book.

1.1 Organizing data. Let the reader consider the following sets of simple data. They will be referred to later by the four letters, A, B, C, D which precede them.

A: A die is thrown five times, and the scores are 3, 2, 3, 1, 6.

B: A stone is thrown up in the air, and its observed heights (in metres) after 1, 2, 3, 4, 5 seconds are 17, 24, 21, 8, -15. (It is assumed that the stone is free to fall *below* the points of projection, so that after 5 seconds it has fallen 15 metres below it.)

C: For five numbers 1, 2, 3, 4, 5 one is added to their cubes, giving 2, 9, 28, 65, 126.

D: The probabilities of throwing exactly two heads in 2, 3, 4, 5, 6 tosses of a coin are calculated, and are found to be $\frac{1}{4}, \frac{3}{8}, \frac{3}{8}, \frac{5}{16}, \frac{15}{64}$.

Now in each of these we have sets of numerical data related to some physical or mathematical situation. The first question we must ask is: how can such data be best organized; in what patterns can it be arranged?

1.2 Function, Domain, Range, Image. The most obvious point about the examples we have given is that, in each case, or certainly in B, C, D, we have two sets of numbers closely related.

Thus in B the set $\{1, 2, 3, 4, 5\}$ is related to the set $\{17, 24, 21, 8, -15\}$; the numbers in the first set measuring times (in seconds); those in the second set measuring distances (in metres).

In D we have the set $\{2, 3, 4, 5, 6\}$ related to the set $\{\frac{1}{4}, \frac{3}{8}, \frac{5}{16}, \frac{15}{64}\}$. Again it is easy to see what the numbers measure.

Even in the first example, A, we have related sets. The first set could be the set of throws, which we could denote by the letters T_1, T_2, T_3, T_4, T_5. Thus we could take the related sets $\{T_1, T_2, T_3, T_4, T_5\}$ and $\{3, 2, 1, 6\}$. Alternatively we could use the numbers 1, 2, 3, 4, 5 to *measure* the throws (just as in *B*, 1, 2, 3, 4, 5 measure five lengths of time spent in the air); so that we would have the related sets

$$\{1, 2, 3, 4, 5\} \quad \text{and} \quad \{3, 2, 1, 6\}.$$

In this book we shall use sets of numbers in this way: numbers which may well measure something in a physical problem (such as in *A*, *B*, *D*), rather than just be numbers in their own right (as in *C*). To understand how such sets effect a physical problem we must, of course, understand what they measure; but in much of the analysis we shall simply be investigating the relationship between the sets themselves. What can we now say about such relationsips? Again, it is very obvious that there is a relationship between the individual elements of each pair of sets. To each element of the first set there corresponds a definite element of the second.

This in *B* can easily be shown by the table:

$$\begin{array}{ccr} 1 & \to & 17 \\ 2 & \to & 24 \\ 3 & \to & 21 \\ 4 & \to & 8 \\ 5 & \to & -15 \end{array}$$

After 3 seconds the stone has risen 21 metres. Therefore for this particular example the element 3 in the first set is firmly related to the element 21 in the second.

Thus the information can be organized into a set of ordered pairs: $\{(1, 17), (2, 24), (3, 21), (4, 8), (5, -15)\}$. Sets of this type we will call *functions*, though we will find it more convenient in this chapter to write them in the unconventional form:

$$\left\{\begin{matrix} 1 & 17 \\ 2 & 24 \\ 3 & 21 \\ 4 & 8 \\ 5 & -15 \end{matrix}\right\}.$$

Thus from *C*, for instance, we derive the function

$$\left\{\begin{matrix} 1 & 2 \\ 2 & 9 \\ 3 & 28 \\ 4 & 65 \\ 5 & 126 \end{matrix}\right\}.$$

In a function then, we see that to each element of the first set there corresponds a definite element of the second. We might be tempted, from the example in *B*, to go on and reverse this proposition: to say 'to each element of the second set there corresponds a definite element of the first'. This is true enough in B. There is, for instance, only one length of time after which the stone has risen 24 metres, and that is 2 seconds; therefore, we might say, to the element 24 of the second set there corresponds the element 2 of the first.

But this is not always true. Take A for instance. The result 3 appears in both the first and third throw. Therefore, while to the element 1 of the first set there certainly corresponds the element 3 of the second, to the element 3 of the second there corresponds *both* 1 and 3 of the first. We illustrate this in Figure 1.

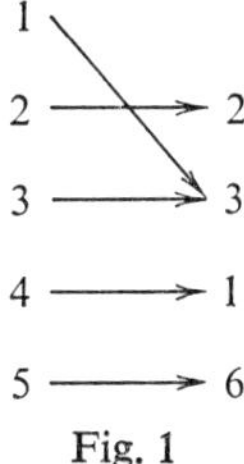

Fig. 1

A function then embodies a lopsided relationship. While to each element of the first set there corresponds a unique element in the second, the reverse is not true. To one element of the second there may correspond more than one in the first.

It is important therefore to be clear about which is the first set and which the second. With this in mind we may use the phrase '4 *maps onto* 1' to indicate one of the correspondences in this example, noting that the 4 comes from the first set and the 1 from the second. We speak then of elements of the first set mapping onto those of the second.

Hence, in a function, more than one element of the first set may map onto the same element of the second, as we have seen in A; and likewise in D where 3 and 4 in the first set both map onto $\frac{3}{8}$ in the second.

We distinguish the two sets by calling the first the *domain* of the function and the second its *range*.

Thus we see our language developing. We speak of the *domain* and *range* of a *function*. To each element of the domain there corresponds just one of the range; or each element of the domain *maps* onto just one of the range; and we speak therefore in any function of the *domain mapping onto the range*—not the other way round.

Consider a college dance. The boys and girls are kept rigorously apart (by over zealous organizers) until the music is about to begin. Then, naturally enough, each boy rushes wildly and grabs the girl whom he fancies most. (Of course in practice this picture might be complicated by corresponding initiatives in the opposite direction, but this, being studious of clarity, we will neglect: this process is called 'mathematical modelling';

see Chapter 4.) Each boy clings determinedly to the girl he first grabs, so that some girls are entangled with two or three boys and some without partners at all.

Could a function—in the sense in which we have been using the word—be extracted from this situation?

The answer is that it could, after a little sorting out.

First we shall give all the boys and girls numbers, so that we can use sets of numbers to indicate them and their relationship. What the numbers are does not matter; so long as no two boys and no two girls have the same number.

Let us suppose that the results of the rush are shown in Figure 2.

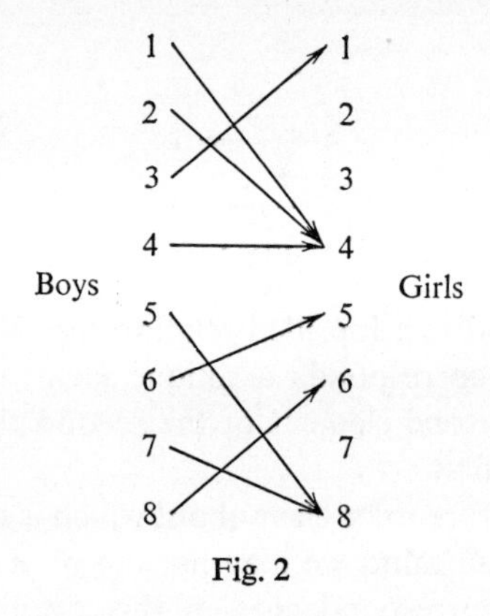

Fig. 2

The domain consists of all the boys (or rather, all that rush). The range, however, consists only of the girls who get partners; since—see Figure 2—girls 2, 3, 7 have no boys corresponding to them they cannot be members of the range.

When the relative size of the numbers in the two sets measures something of interest, it may be helpful to illustrate the function (or, as we may say, draw a graph of the function) by displaying the domain and range sets on number lines, as for instance with data D in Figure 3.

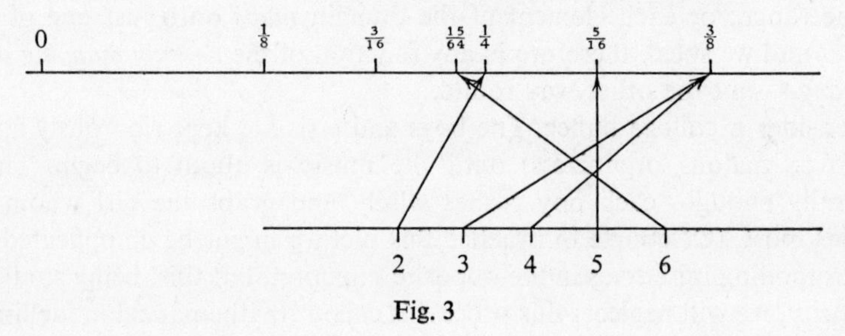

Fig. 3

We can see here that as we consider progressively larger elements of the domain, the elements of the range into which they map rise to a maximum and then decrease.

As an alternative way of saying that 'a maps into b', we shall call b the *image* of a. So in our present example we could say, for instance, that the image of 2 is $\frac{1}{4}$.

Exercise A

1 For each of the following sets of data, inventing numerical codes where necessary, (i) write out a function, (ii) state its domain and range in set notation, (iii) illustrate the function using arrows to connect corresponding elements:

(*a*) a balloon is blown up, and when its lengths are 2, 3, 4, 5, 6 cm, its corresponding volumes are 3, 10, 18, 42, 75 cm^3;

(*b*) a space capsule at noon on the 2nd, 3rd, 5th, 6th day after blast-off is 210000, 340000, 340000, 133000 km from the earth's centre;

(*c*) the telephone bills for four people are £14, £8.50, £21, £17.75.

2. Illustrate the function $\left\{\begin{matrix} 3 & 7 \\ 4 & 10 \\ 5 & 13 \\ 6 & 16 \end{matrix}\right\}$ as in Figure 3. (i) What does 3 map onto? (ii) What maps onto 10? (iii) What is the image of 5?

3. For the function $\left\{\begin{matrix} 3 & 5 \\ -4 & 2 \\ 6 & 5 \\ 2 & 10 \end{matrix}\right\}$ state (i) the image of 3; (ii) the elements of the domain which map onto 5; (iii) the image of 2; (iv) the range.

4. If the domain of a function is $\{-2, -1, 0, 1, 2\}$ and the image of each domain element is its square, what is the range? What numbers map onto 4? What does -1 map onto? Illustrate the function as in Figure 3.

5. Write down three different functions whose domains are $\{2, 3, -5\}$ and whose ranges are $\{-7, 17\}$. How many such functions are there in all?

6. Write down the domains and ranges of the following functions:

$$\text{(i)} \left\{\begin{matrix} 1 & 2 \\ 3 & -1 \\ 6 & 2 \end{matrix}\right\}; \quad \text{(ii)} \left\{\begin{matrix} 3 & -1 \\ 1 & 2 \\ 6 & 2 \end{matrix}\right\}; \quad \text{(iii)} \left\{\begin{matrix} 1 & 2 \\ 6 & -1 \\ 3 & 2 \end{matrix}\right\};$$

$$\text{(iv)} \left\{\begin{matrix} 6 & 2 \\ 1 & 2 \\ 3 & -1 \end{matrix}\right\}; \quad \text{(v)} \left\{\begin{matrix} 6 & 2 \\ 3 & 2 \\ 1 & -1 \end{matrix}\right\}.$$

How many different functions are there?

1.3 Functions as mathematical elements. Since a function, like a matrix for instance, is a single mathematical element, we shall denote it when convenient by a single letter. As an example, consider

$$f = \begin{Bmatrix} 1 & 7 \\ 2 & 9 \\ 3 & 7 \\ 5 & -3 \\ 8 & -10 \end{Bmatrix}.$$

Under this function (as the phrase goes) the image of 3 is 7. There is a mathematical process by which we can obtain the number 7 from the function f and the number 3. This may be written

$$\begin{Bmatrix} 1 & 7 \\ 2 & 9 \\ 3 & 7 \\ 5 & -3 \\ 8 & -10 \end{Bmatrix} (3) = 7.$$

The process here is a simple one of selecting from the function the number which corresponds to 3. (A similar situation in principle arises when a position vector is transformed by a matrix into a new position vector; then the process involved is called premultiplying the original position vector by the matrix.)

We can rewrite the statement

$$f(3) = 7$$

and say 'the image of 3, under the function f, is 7'. Similarly, $f(1) = 7$ and $f(2) = 9$.

If we had a different function, let us say the function whose domain is $\{2, 4\}$ and range $\{72\}$, we could denote it by a new letter, g for instance, and write $g(2) = g(4) = 72$.

1.4 Variable. In Section 1.1 we had the data C where the elements 1, 2, 3, 4, 5 mapped onto 2, 9, 28, 65, 126. Here there was a definite mathematical law for calculating the image of any element of the domain: cube the element and add one. Thus

$$f(1) = 1^3+1, \quad f(2) = 2^3+1, \quad f(3) = 3^3+1,$$

and so on. Now these equations powerfully suggest the possibility of *generalization* for this particular function. Clearly it can be tedious writing out these results at such length, particularly where the domain is a big one. However we can introduce a symbol to do duty for them all. The symbol usually employed is x.

We can write $$f(x) = x^3+1,$$
where x *is a symbol which may be replaced at will by any element of the domain.*

We call such a symbol a *variable* or *place-holder*; variable most commonly, though the name is rather a misleading one. It does not mean a number which can mysteriously change its size. It is not strictly a number at all. It is a symbol which can be replaced by a number (any element of a set, in this case the domain) to give a true statement.

Thus, '$f(x) = x^3+1$' means 'if we replace x by any element of f's domain, then $f(x) = x^3+1$'.

So for instance, since 4 is an element of the domain, it follows that

$$f(4) = 4^3+1 = 64+1 = 65.$$

The symbolic notation is essentially a shorthand rule for telling us the image of each element of the domain.

Some of the expressions used with variables may foster the notion that variables are numbers which can mysteriously change size; that they can be blown up and deflated like a balloon. There is no harm in them, so long as we understand clearly what they mean. Thus for instance 'as x gets larger, $f(x)$ gets larger' really means 'if we replace x by a sequence of numbers each larger than the previous one, we will get a sequence of images each larger than the previous one'.

And this is true in the particular case where $f(x) = x^3+1$. If we replace x by the series 1, 2, 3, 4, 5 we get the series of images 2, 9, 28, 65, 126.

Likewise: 'If $x > 2$, $f(x) > 10$' means 'if we replace x by an element of the domain greater than 2, we will find that its image is greater than 10'. This also is true.

We will find other shorthand expressions connected with variables; but the reader should soon have no difficulty in sorting them out for himself.

1.5 Function of variable. We can think of $f(4) = 65$, for instance, as saying that 65 is the result of combining the function f and the number 4 in the way defined above. We use the word 'of' to describe this operation, so the equation reads 'f of 4 equals 65' (as an alternative to 'the image of 4 under the function f is 65').

Now if $f(x) = x^3+1$, it is a common and convenient practice to refer to x^3+1 as *a function of* x. The reader may like to link this in his mind with the reading of $f(x)$ as 'f of x' which is close to 'function of x'.

We note that a 'function of x' is a variable, not a number. It is a symbol in which x can be replaced by a number to get its image under the function.

The phraseology in calculus has not grown systematically but certain usages have proved their worth and we shall take advantage of them. There is one such phrase which is most useful.

In terms of data B, suppose the function involved is f. Then when the time is x seconds, the height is $f(x)$ metres; and we say: '*height is a function of time*'.

If we make a statement like 'the height (in metres) is seven times the square of the time (in minutes)', the symbols 'height' and 'time' can be replaced by numbers to give a statement which is true if the original statement was true. 'Height' and 'time' (or strictly 'the height' and 'the time') are therefore *variables* as we earlier defined them. So here we see that indeed 'height is a function of time'.

Let us follow this through in reverse.

Suppose we are thinking about the volumes of spheres of various radii; we can say 'volume is a function of the radius'. What this means is that there is a function—call it V—which maps the number of units in the radius, whatever it may be, onto the corresponding number of units in the volume.

In fact $V(r) = \frac{4}{3}\pi r^3$.

Even if we did not know the formula for the volume in terms of the radius, we might still say 'volume is a function of the radius', but then this would merely be an expression of the realization that to each radius there corresponds a unique volume. We often use such statements in this loose but expressive way.

Exercise B

1. Find where possible

(i) $\begin{Bmatrix} 2 & 1 \\ 5 & 0 \\ 7 & 3 \end{Bmatrix}(5)$; (ii) $\begin{Bmatrix} 3 & 4 \\ 2 & 3 \end{Bmatrix}(3)$; (iii) $\begin{Bmatrix} 2 & 1 \\ 0 & -4 \\ -1 & 7 \end{Bmatrix}(0)$; (iv) $\begin{Bmatrix} 4 & 3 \\ 2 & 1 \\ 1 & 2 \end{Bmatrix}(3)$.

2. If
$$f = \begin{Bmatrix} -1 & 6 \\ 0 & 10 \\ 1 & 15 \\ 2 & 30 \end{Bmatrix} \text{ and } g = \begin{Bmatrix} 2 & 4 \\ 7 & 11 \\ -3 & 4 \\ 4 & 7 \end{Bmatrix}$$
what are (i) $f(2)$; (ii) $g(-3)$; (iii) $f(0)$; (iv) $g(4)$; (v) $g(7)$? What is the range of g? What does 2 map onto under f and under g? Solve for a, $g(2+a) = 4$.

3. If $f(y) = 2y-4$, what is $f(3)$? If $f(y) = 6$, what is y?

4. If $f(x) = x^2$ and the domain of f is $\{-2, -1, 0, 1, 2\}$, what is the range of f? What is $f(-1)$? If $f(x) = 4$, what is x?

Illustrate the function as in Figure 3.

5. For the function $f = \begin{Bmatrix} 0 & 3 \\ 1 & 5 \\ 2 & 7 \\ 3 & 9 \\ 4 & 11 \\ . & . \\ . & . \end{Bmatrix}$ whose domain is the set of all non-negative integers, write down in shorthand (i) 'the image of 2 is 7'; (ii) '4 maps onto 11'. Complete the statement '$f(x) = ...$'.

6. Complete the statement '$g(x) = ...$' for the function $g = \begin{pmatrix} 1 & \frac{1}{2} \\ 2 & \frac{1}{3} \\ 3 & \frac{1}{4} \\ 4 & \frac{1}{5} \\ \cdot & \cdot \\ \cdot & \cdot \end{pmatrix}$ whose domain is the set of all positive integers.

What is $g(500)$? If $g(x) = 1/100$, what is x? Does $g(0)$ have a meaning? Can $g(x) = 0$?

7. Explain carefully what is meant by (i) 'as x increases, $f(x)$ decreases'; (ii) 'when $x = 0, f(x) = 5$'; (iii) 'when x is doubled, $f(x)$ is trebled'.

8. What is the meaning of:

(i) temperature is a function of time, for a cooling body?

(ii) distance is a function of time for a man walking at 4 km/h? After t hours, what function of t is the number of kilometres walked?

9. For a solid circular cylinder of volume 10 cm^3, express the total surface area as a function of the radius.

2. VARIETY IN FUNCTIONS

2.1 Discrete and continuous variables. In the examples in Section 1.1, we choose domains limited in each case to five integers. But this choice was not, of course, imposed on us by the physical situations behind A, B and D; nor the mathematical problem behind C. In each case we could have chosen domains as big as we liked: in B we could, with suitable equipment, have recorded more heights, both between those recorded in Section 1.1, and *after* five seconds had elapsed. In D we could have worked out probabilities of two heads in 7, 8, 9, ..., throws. And in A we could have gone on throwing the die and recording the scores, and in C gone on cubing numbers and adding one, indefinitely.

Nevertheless there is one important difference between A and D on the one hand and B and C on the other.

C shows it most clearly. There we could have extended the domain to include *all* numbers, or all numbers between certain limits. For instance we could have defined the function to be given by the formula $f(x) = x^3+1$ for all x (say) between -1 and 6. There would have been infinitely many elements of this domain, since there are infinitely many numbers between -1 and 6. But there is no number in this interval (or anywhere else) to whose cube we cannot add one.

Likewise in B we can at least *imagine* the height of the stone after any length of time. In practice, with the limitations of our instruments, we would be restricted in the measurements we took, as to their accuracy. But there seems to be no theoretical reason why we could not indefinitely extend the number of elements in the domain, and—more to the point—crowd them together as close as we like.

In D on the other hand, with our system of measurement, elements in the domain can only be whole numbers.

Likewise in A. Of course in A the system of measurement is a little arbitrary. There is no reason why we should not measure the successive throws by the numbers 0, 0·1, 0·2, 0·3, ..., thus getting elements in the domain more crowded together. But it is hard to imagine a system of measurement in which we do not go through the domain in a series of finite hops.

This points to an essential difference. We shall describe a domain through which we can move smoothly and steadily without jerks as a continuous domain, and the associated variable as a continuous variable. One in which we must move in finite hops we shall call a discrete domain. We shall not attempt to define these terms more precisely here.

We shall mostly—but not always—be concerned with continuous domains in this book.

2.2 Formula functions. There is another important difference, or series of differences, illustrated by examples A, B, C, D. Let us consider them in the order C, D, B, A. The point at issue is the possibility of finding a formula from which we can find the image of any element in the domain of a function.

In C the formula is virtually given. We know, if f is the associated function, whatever the domain: $f(x) = x^3+1$. This formula allows us to calculate the image, no matter of what element.

The same is true in D. Here we have to calculate a formula, but as we originally calculated the probabilities, this should afford no difficulty. We are merely generalizing our previous work.

If we toss the coin x times, two heads can fall in $\frac{1}{2}x(x-1)$ positions. The probability of each of these positions is $\frac{1}{2}^x$. Hence the probability of two heads in x tosses is

$$x(x-1)\,(\tfrac{1}{2})^{x+1}.$$

So we have to deal with the function f where

$$f(x) = x(x-1)\,(\tfrac{1}{2})^{x+1}$$

over the domain of all positive numbers (including 1, since $f(1) = 0$, which is what the formula gives).

But what of B?

For here our data was experimental, and it is hard to see what part a formula such as those we have been dealing with can play with experimental data. Suppose we hit on a formula, by luck say, which happened to fit the five pairs of numbers we have given, would this be 'the' formula? Surely we should object that better measuring instruments would have shown up errors in the numbers we gave and that the formula must therefore be wrong.

Nevertheless we do not doubt that there is a height at any time, and we would expect the height to change in some regular pattern. In fact it is not hard to guess, looking at the figures in B, that the height after 1·5 seconds is between 17 and 24 metres, and somewhat more than half way between them. We might indeed expect there to be pattern in the images, but one very reluctant to adapt itself to a neat algebraic formula.

Both expectations would be right. As it happens there is a simple formula for this data: $f(x) = 22x - 5x^2$, as the reader can easily verify. This is so because, in spite of the mendacious formulation of B, the data was doctored to agree with this fomula. This would not happen in real life: we are unlikely to find a right formula for such a problem. However, mathematicians do spend a great deal of time trying to find approximate formulae to fit such data; and ways of doing this, how to fit mathematics to the physical world, will be one of the major problems of this book.

So we can assert that although the experimental data may not be bound to a formula, we can often find a formula which will fit it approximately.

But how about A?

It is possible to fit a formula to the scores obtained in the five throws, but it would be practically valueless. It could not be used to predict what would happen on a subsequent throw, nor if the experiment were repeated could it be used again. There can be no formula pattern behind these random results; though with such results, patterns of a different sort can be associated.

We introduced this type of example only so that we could now exclude it from further consideration, in order to define more clearly the nature of this book. We shall be concerned with what may be called *formulae functions:* functions which show a pattern; functions in which, even if we find it hard to fit a formula, we can at least imagine a formula.

It is the relation between functions and the formulae defining them, and between physical problems and the associated formulae we derive, that is our study here.

But in order to find functions relevant to physical problems, we must study functions first at their driest and most abstract: functions as in C which are defined by a formula. When we have discovered how these behave, we are better placed to consider which is the most likely candidate in a physical problem.

2.3 Behaviour of functions. Both the behaviour of functions and the nature of formula functions are best illustrated by drawing their graphs.

We have seen so far three ways of illustrating a function: the table (p. 2); the arrow diagram (Figures 1 and 2); and the parallel number lines (Figure 3). None of these is of much use.

The standard way of illustrating the 'behaviour' of a function is, of course, displaying the relationship against two perpendicular number lines.

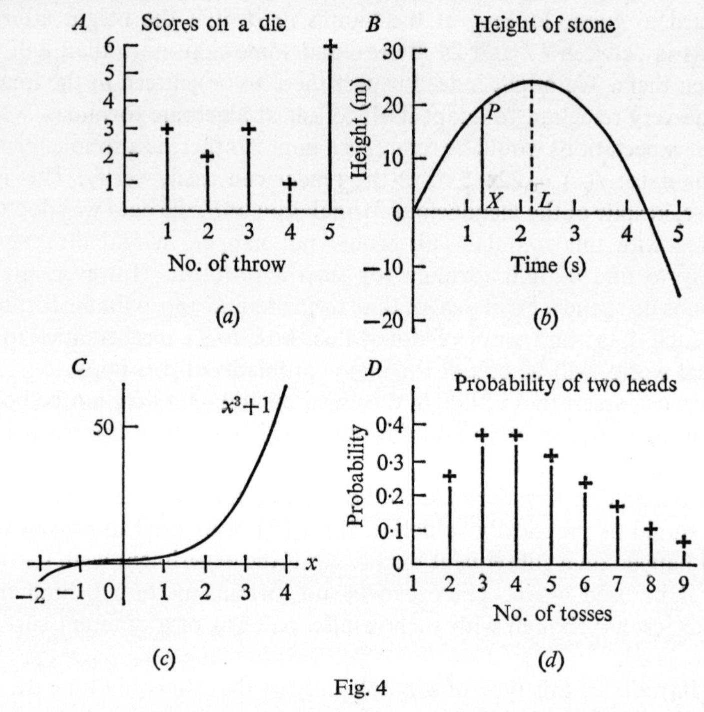

Fig. 4

If, for instance, the image of the domain element 1 is 3, denote this by erecting a line of length 3 directly above the point marking 1 on the horizontal number line, the *domain axis.*

Figures 4(*a*), (*b*), (*c*), (*d*) illustrate this for the examples *A*, *B*, *C*, *D*; but the upright lines have been left out for *B* and *C*, and their domains have been extended to cover all numbers between 0 and 5 for *B*, and -2 and 4 for *C*.

Certain distinctions are clear: those in fact that we have already made. *A* and *D*, with discrete variables, proceed by a series of hops: the graph being a succession of isolated points, those at the top of the lines. But whereas in *A* the points are incoherently distributed (as befits the random nature of the data), in *D* they clearly follow some sort of pattern: they suggest in fact an orderly rise and fall, and even that some smooth curve could be found to pass through them. This is the characteristic of a fomula function with discrete domain.

B and *C* on the other hand have formulae functions with continuous domains, so the smooth curve is already manifest in the graphs, and the behaviour of the function is clearly seen.

Take B, for example.

Here as the variable increases, its image increases till it passes through a maximum corresponding to L in the figure. (We see this by considering the growth and then diminution of the line PX as we imagine it moved smoothly to the right, with X moving along the domain axis.) Thereafter it falls to zero, and afterwards becomes negative, falling ever more steeply as it goes on.

We might translate this into physical terms by saying that as the stone is thrown up it slows down until it reaches its highest point (after time measured by OL). Thereafter it falls back towards the point of projection, and on past this point (so that it is a negative height above it) getting always faster as it falls.

This is mere commonsense. But it also illustrates what we mean by the 'behaviour' of a function; and also the physical interpretation of such behaviour. Much more will be said about this later in the book.

In investigating the behaviour of a function, we may want answers to such questions as:

as x increases, does its image increase?
if so, does it rise to a maximum?
for what x, and how big is the maximum?
if the image increases as x increases, how sharply does it increase?
as x increases, does the image increase more sharply or less sharply?
or does it change from one to the other, and if so where?
does the image ever suddenly move off to infinity?
what does the image do as x moves to infinity? Go to infinity too or settle down to some finite value? or oscillate wildly all over the place?
when is the image zero?
for what values of x is it positive, and for what negative?
is this an excitable function which does strange things?

The question about how sharply the image rises with x is worth looking at in connection with B. It is easy to see from the data in Section 1.1, that in the first second it rises 17 m; in the second 7 m; in the third *falls* 3 m; in the fourth falls 13 m; in the fifth falls 23 m.

It is not hard then to see that height rises sharply with time to begin with; that thereafter it slows down; that when it falls, it falls ever more sharply as time goes on.

But how is this reflected in the shape of the graph? This question will return later; but it may be worth the reader considering it, briefly, now.

Let us say again that it is the 'behaviour' of functions and how this can be interpreted physically that is our chief concern.

Exercise C

1. The time taken to cross the Atlantic (5000 km) is a function of the speed. If the time is $f(x)$ hours, at a speed of x km/h, express $f(x)$ as a formula in x.

Sketch the graph of f for values of x from 50 to 5000.

Is x a continuous or discrete variable?

Describe the behaviour of f, if the domain extends over all positive numbers. Interpret the behaviour physically.

2. The number of different ways in which we can choose r cabinet ministers from a cabinet of ten, for different values of r, is given by the following table:

r:	1	2	3	4	5	6	7	8	9
No. of ways:	10	45	120	210	252	210	120	45	10

Complete the sentence: 'the number of different ways is a function of ...'. Sketch the graph of this function. Is r a continuous or discrete variable? Describe the behaviour of the function and interpret it physically.

3. Six dice are thrown together. The probabilities of getting 0, 1, 2, 3, 4, 5, 6 sixes in a throw are 0·335, 0·402, 0·210, 0·053, 0·008, 0·001, 0·000 (to 3 d.p.) respectively.

Complete the sentence: 'the probability is a function of ...'. Sketch the graph of the function, describe its behaviour and interpret this physically.

4. For a spreading ink blot, we can say that area is a function of the time; area is a function of the radius; and radius is a function of the time. Denote these three functions by f, g, h respectively. In a mathematical model of the event we shall suppose the blot is always circular; that initially its area is 4 cm^2 and that its area increases steadily at 2 cm^2/s for the first 5 seconds.

Draw the graph of f and the graph of g (where $g(r) = \pi r^2$). By taking readings from these graphs, draw the graph of h. Describe the behaviour of each of f, g, h and interpret in physical terms.

3. PHYSICAL PROBLEMS

We shall finish this chapter by trying to show how functions arise in simple physical problems, and how their 'behaviour' is reflected in the nature of the problems. In many of these it will be possible to calculate (by simple geometry, or the theory of probability) formulae appropriate to the functions; and we shall return to this matter in Chapter 4. Here however we shall content ourselves with an entirely qualitative approach. We shall ask such questions as: when some quantity (height, probability, time, for instance) increases, what happens to some other quantity dependent on it? It is this dependence of one quantity on another that suggests the presence of a function.

Example. A right circular cylinder fits tightly inside a fixed hollow sphere of radius 10 cm. How will the cylinder's volume change as its shape changes?

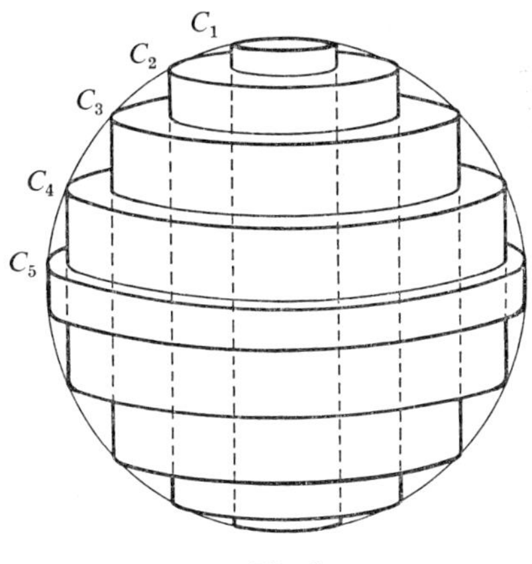

Fig. 5

Now clearly in this problem we can find cylinders of widely differing shape (see Figure 5). We can have long thin cylinders, almost like pencils as in C_1, stretching along a diameter. Or we can have wide flat cylinders like discs, as in C_5, spanning a diametral section. Or we can have cylinders of various shapes like C_2, C_3, C_4 in between.

Now—to think functionally—we want a variable to measure the different sorts of cylinder we can get. We can think of the volume as being a function of either the height of the cylinder or of its radius, or indeed of other less obvious variables. We shall choose the radius of the cylinder, measured in centimetres, so that the variable is r, say. Replace r by 5, and we have to deal with a cylinder of radius 5 cm.

A little thought convinces us that there can only be one cylinder of radius 5 cm fitting tightly inside the sphere, something like C_3, though of course we can rotate it into different positions.

Again as we consider the sequence of cylinders from C_1 to C_5, clearly r increases from something a little more than zero to something a little less than 10. And evidently we should include zero and 10 in the domain, since at one extreme there will be a 'cylinder' consisting of a single line; and at the other of a circular plane, though the volumes of both will be zero.

This gives us the best idea of a function arising from the problem. Evidently we have a function mapping the number measuring the radius (in cm) onto the number measuring the volume (in cm^3). Under this function the images of both 0 and 10 will be 0; but between these limits the images will be positive and will, presumably, rise to a maximum for a cylinder something like C_3. This is reinforced by our instinctive sense that C_1 and C_5 have small volumes, but that those in the middle are more substantial.

We summarize this information in the graph of the function shown roughly sketched in Figure 6. Note that we should not conclude that the maximum is exactly half way, where $r = 5$. In fact it is not; though at the moment we have no means of locating it exactly.

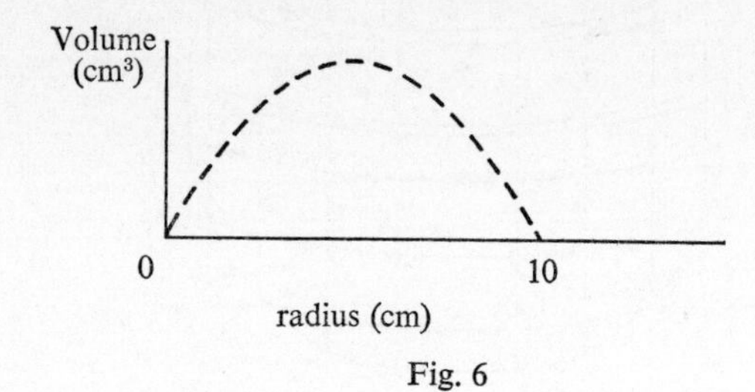

Fig. 6

Answer then: as the cylinder changes from being a long thin cylinder to a normal cylinder to a short fat cylinder, its volume rises from zero, passes through a maximum, and falls to zero again. We do not know where the maximum is.

But this information is best contained in a rough sketch of the function's graph.

Exercise D

1. A space capsule enters the earth's atmosphere. Explain the meaning of: 'the surface temperature of its heat-shield is a function of the time'; inventing numerical data if it helps to make your meaning clear.

Sketch the graph of this function.

Choose another variable of which the temperature is a function and sketch the graph of this function.

2. A metal ball at $-100°$ C is placed in a large volume of water at $10°$ C. State, in general terms, a function relevant to this situation and sketch its graph.

3. The range of a shell is a function of the angle of elevation of the gun. Sketch the graph of this function.

The range is also a function of the greatest height which the shell reaches. Sketch the graph of this function.

Is the greatest height a function of the range?

4. A straight wire touches a disc at the point U, as in Figure 7. A, B are points of the wire at the same distance, x cm, from U. A taut piece of string, of length l cm, runs from A to B around the disc, leaving it at the point T. Angle $UOT = \theta°$.

Sketch the graphs of functions f and g, where $l = f(x)$ and $l = g(\theta)$.

5. Two rods, each of length l metres, lie in a vertical plane and are joined together at one end (see Figure 8). Each rod passes through a small fixed ring, one at P and the other at Q, and their lower ends rest on the horizontal ground. The whole configuration is symmetrical with the rods inclined at $\theta°$ to the horizontal and their common point x metres above the ground.

Sketch the graphs of the functions f and g, where $l = f(x)$ and $l = g(\theta)$.

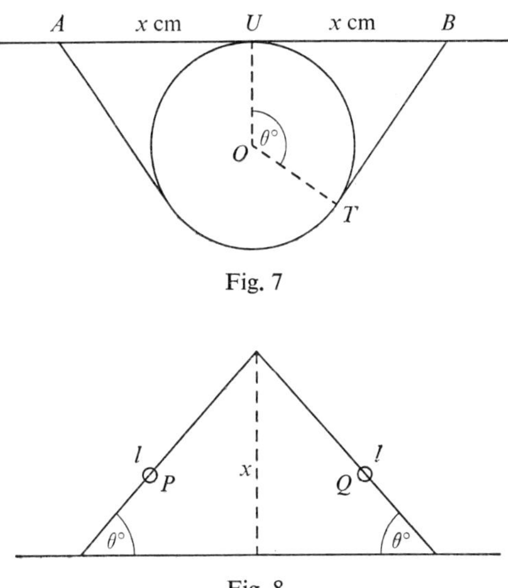

Fig. 7

Fig. 8

6. A half-kilogram of butter is made into different cuboid shapes, each having a square cross-section. Choose a domain variable and sketch a graph of the total surface area as a function of this variable to show how this area varies with the shape.

7. Different conical tents each have a curved surface area of $20\,\text{m}^2$. Draw a graph to show how their volumes vary with their shapes.

8. The probability that, in a random group of people, at least two of them have their birthday anniversary on the same day, is a function of the number of people in the group. Sketch the graph of this function.

If the words 'at least' are replaced by the word 'exactly', sketch the graph of the new function.

9. An aircraft flying at a constant height and steady course passes directly above a ground observation telescope. The *rate* of rotation of the telescope in a vertical plane is a function of (i) the time, (ii) the angle of elevation of the telescope. Sketch graphs of these two functions for the journey from horizon to horizon (where in (ii) the angle is measured from 0 to 180°).

10. On a windy day the bell in a bell tower clangs in a random sort of way, on average over a long period about once every minute. The probability that it clangs exactly once in any interval of t minutes can be regarded as a function of t. Sketch the graph of this function.

Sketch in the same figure the graphs of the probability that it does not clang in an interval of t minutes, and that it clangs exactly twice in an interval of t minutes.

2

POLYNOMIALS AND RATIONAL FUNCTIONS

In Chapter 1 we discussed what was meant by the behaviour of a function, and saw that we could most easily determine a function's behaviour by sketching its graph. We turn now to some simple techniques for graph sketching, and we shall apply them in this chapter to *polynomial* and *rational* functions: functions, that is, which can be built up from integral powers of a variable, like the functions f such that

$$f(x) = 3x^7 \quad \text{or} \quad x^5 - 2x^2 + \frac{1}{x} \quad \text{or} \quad \frac{x}{2x^2+1}.$$

They are techniques however which we shall use throughout the book.

1. LINEAR FUNCTIONS

1.1 The simplest functions of this kind we call *linear functions.* The nature of a linear function is suggested by the following two problems.

(*a*) A rectangular tank contains water to a height of 10 cm and is filling up at the steady rate of 3 cm/h. How does the height vary with time?

(*b*) One end of a poker is plunged into a fire, and its temperature falls off steadily with distance from that end. 7 cm from the end it is 80 °C; at 17 cm it is 60 °C. What is the temperature at the far end if the poker is 30 cm long?

In these we think naturally of the height as a function of time; and of temperature of the poker as a function of the distance from the end plunged in the fire.

What are the common characteristics of these functions?

(The reader, if he is unfamiliar with this work, should draw accurate graphs of height of water against time; and of temperature against distance; and check the observations below against his graphs.)

Most obvious is the *steady rise or fall of the image with respect to the domain variable:* the rise in height with respect to time; the fall of temperature with distance.

We may put this by saying that *equal changes in the domain variable product equal changes in its image:* for instance, equal changes in distance down the poker produce equal falls in temperature; and this leads us to suppose that the graphs are straight lines (hence *linear* functions).

Or again we may note that *change in the image is proportional to change in the domain variable:* change in height (in cm) is 3 times the number of hours in which it occurs—here then the constant of proportionality is 3 (measured in cm/h); and this constant appears as the gradient of the graph. It measures the 'rate of rise'. (What is it, and what does it measure in the second case?)

Lastly the formula. This can easily be calculated in the first problem. If the height after t hours is $H(t)$ cm, then—since in t hours the rise is $3t$ cm—

$$H(t) = 10+3t.$$

This suggests that a linear function will be defined by a formula of the type

$$f(x) = a+bx,$$

where a and b are constants; and we shall see in fact that functions defined in this way have all the characteristics we have mentioned.

1.2 $f(x) = a+bx$. Let us take such a function, say f, where $f(x) = 3+2x$ and sketch its graph.

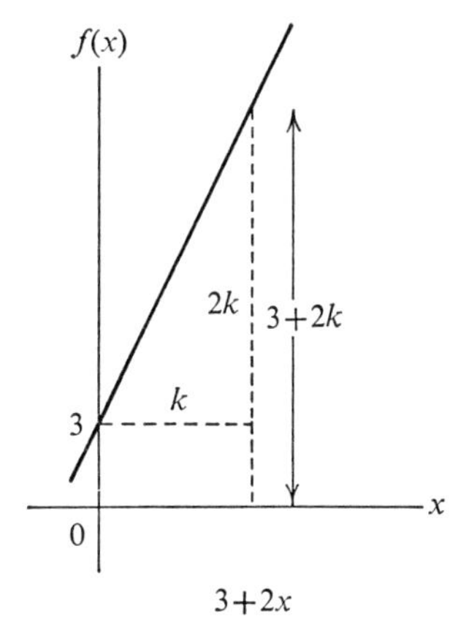

Fig. 1

Under this function the image of 0 is 3, giving the point (0, 3) on the graph. The images of 1, 2, 3 are 5, 7, 9, giving the points (1, 5), (2, 7), (3, 9). It seems from this that whenever we increase x by 1, we increase its image by 2, so that the points all lie on the line through (0, 3) having gradient 2.

To prove it in general however we note that k maps onto $3+2k$; so that increasing x from 0 to k, we increase its image from 3 to $3+2k$, that is by $2k$ (see Figure 1). Hence this point too lies on the line through (0, 3) with gradient 2.

Hence the entire graph is this line.

A similar argument could be used for something like $f(x) = 5-\frac{1}{2}x$. Here 0 maps onto 5, and the graph passes through (0, 5).

Increase x by k from 0 to k, and we increase its image from 5 to $5-\frac{1}{2}k$: that is, we *decrease* it by $\frac{1}{2}k$ (see Figure 2). The gradient of the line joining P and Q is hence $-\frac{1}{2}$, and we conclude that the function's graph is a line through (0, 5) with gradient $-\frac{1}{2}$.

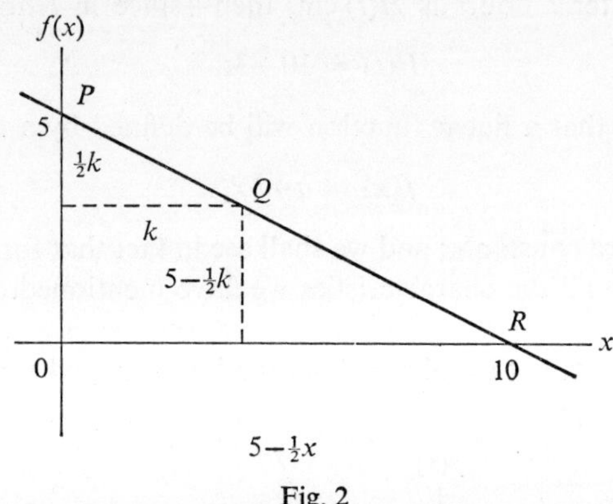

Fig. 2

We note also that, since it has this gradient, OR is twice OP, that is 10. Hence the graph crosses the x-axis at (10, 0): an answer we could have got from considering what value of the domain maps onto zero:

$$\text{if} \quad f(x) = 5-\tfrac{1}{2}x = 0, \quad \text{then } x = 10.$$

1.3 Let us finish by returning to problem (*b*) in Section 1.1. Here temperature decreases steadily with distance from the fire, so that it is a linear function of this distance. We can easily solve the problem by common sense. The temperature falls from 80 to 60 °C as the distance increases from 7 to 17 cm: that is, by 20° in 10 cm. Therefore the rate of fall is 2°/cm. The distance from the second point to the end of the poker is 13 cm, since the poker is 30 cm long. Hence in this distance temperature falls a further 26°, bringing it from 60° down to 34°.

It is however instructive to solve this also by using the idea of a linear function.

Let the temperature at x cm from the fire be $T(x)$ °C. Then we know:

$$T(7) = 80, \quad T(17) = 60.$$

But T is a linear function. Hence we can express it in the form

$$T(x) = a+bx.$$

Thus $$T(7) = a+7b = 80$$

and $$T(17) = a+17b = 60.$$

Solving these equations, $b = -2$ and $a = 94$.

Hence $$T(x) = 94-2x.$$

But we wish to find the temperature where $x = 30$.

$$T(30) = 94-2\times 30 = 94-60 = 34.$$

Thus the temperature at the far end is 34 °C.

Exercise A

1. If $f(x) = 4-x$, plot the points on its graph where $x = 0$ and $x = 8$. Hence draw the graph and check its accuracy where $x = 4$.

2. If $f(x) = -2+\frac{2}{3}x$, plot the points on its graph corresponding to the values of $f(0)$ and $f(6)$. Make a table of values of $f(x)$ for integral values of x from -3 to 6 and state its obvious characteristic feature.

3. Using the argument of the second example in Section 1.2, sketch the graphs of:

$$-3+2x; \quad 5-x; \quad -\tfrac{3}{4}x-7.$$

Find their gradients and where they cross the x-axis.

4. Sketch graphs of: $\quad 4x+2; \quad -3-x; \quad 2x.$

Note their gradients and where they cross the x-axis.

5. Draw the graph of f where $f(x) = 3$. What other functions have for their graphs lines parallel to the x-axis?

Is there any function whose graph is a line perpendicular to the x-axis? If so, give an instance. If not, explain why.

6. In the same figure, sketch the graphs of:

$$3-2x; \quad 3-2x+\tfrac{1}{2}x; \quad 3-2x+\tfrac{1}{5}x; \quad 3-2x+\tfrac{1}{10}x.$$

7. $f(x) = a+bx$. $f(1) = 7$ and $f(5) = 19$. Find a and b, and make out a table of values of $f(x)$ for integral values of x from 0 to 5. Explain the significance of b in relation to this table.

8. f is a linear function. $f(-3) = 19$ and $f(3) = 7$. Find the gradient, $f(0)$, $f(-2)$, $f(5)$, and sketch the graph.

9. f is a linear function for which $f(4) = -1$ and $f(8) = 1$. For what value of the domain is the image 0?

10. A man starts 5 km north of a point A, and walks south at 4 km/h for 2 hours. Why is his distance from A a linear function of time? Sketch the graph of this function and find its gradient.

Would his distance from A have been a linear function of time if he had walked west at 4 km/h for 2 hours? Sketch very roughly the graph of whatever function it is.

11. A man walks at a steady speed all afternoon on a straight line through the point A. At 3 p.m. he is 4 km east of A; at 7 p.m. he is 20 km west of A. Express his distance east of A as a linear function of the number of hours that have elapsed since noon and sketch its graph. Where is he at 1 p.m. and at 5 p.m.? What is his speed?

12. The length of the spring of a spring balance is y cm when the mass attached is W grams. If the scale is evenly spaced, what do you deduce about the function which maps y onto W? If $W = 12$ when $y = 25$ and $W = 22$ when $y = 27\frac{1}{2}$, find the length of the spring when it carries no mass. Find also the value of W when $y = 20$ and interpret the result.

13. Two springs are such that $T = \frac{2}{3}x - 12$ and $T = \frac{1}{2}x - 6$, where T newtons is the tension in a spring of length x cm. Is it possible for the tensions to be the same when the lengths are equal to one another?

If so, for what values of T and x? Illustrate graphically.

14. Assuming the formula for the circumference (C) of a circle in terms of the radius (r), sketch the graph of the function which maps r onto C. What is the gradient of this graph? What is the increase in C when r increases from any value by one unit?

Find (i) the difference in lengths of the circuits round a circular track for two men running side by side, 1 m apart;

(ii) the increase in the circumference of the earth if the radius were increased by 7 m.

15. A man walking at 6 km/h covers D km in T hours. Express D as a function of T, and sketch the graph of the function. Sketch in the same figure the graphs of the functions corresponding to speeds of 3 km/h and 8 km/h. What significance would you attach to a negative speed and how would you represent it on the figure?

16. A metal bar is L cm long when its temperature is $C°$ centigrade. L and C are approximately related by the formula $L = 0{\cdot}02C + 300$.

If the temperature increases by 12°, what is the expansion?

If the temperature decreases by 5°, what is the contraction?

17. Over a certain period, the cost of production (£C) of a manufactured article is expressible as a function of the number (n) of articles produced, by the formula $C = a + bn$, where a, b are constants. Find the value of b if an increase of 10 in the number of articles costs an extra £7.50.

2. POWER FUNCTIONS

A power function is that which maps a variable onto some fixed power of the variable: say f, where $f(x) = x^3$ or $f(x) = x^{-5}$.

These occur so frequently that we introduce the symbol P_r to denote the function which maps x onto x^r.

Thus
$$P_2(x) = x^2; \quad P_3(x) = x^3; \quad P_{-1}(x) = x^{-1} = \frac{1}{x}.$$

Note also the trivial cases: $P_1(x) = x$ (a linear function) and $P_0(x) = x^0 = 1$.

In addition we shall use kP_r, where k is a constant, to denote the function which maps x onto kx^r, and $-P_r$ for the function which maps x onto $-x^r$.

2.1 $\boldsymbol{P_2(x) = x^2}$. This is one of the simplest power functions. We will discuss first the form of its graph and then how this can be adapted to get the graphs of some closely related functions.

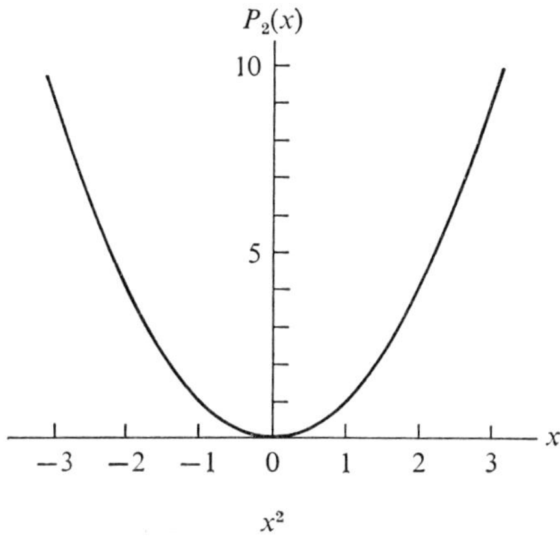

Fig. 3

A table of values is:

x	-4	-3	-2	-1	0	1	2	3	4
x^2	16	9	4	1	0	1	4	9	16

Equal increases in x no longer produce equal increases in the images. x^2 increases by larger and larger amounts as x gets larger by equal amounts.†

x^2 is zero when $x = 0$ and is otherwise positive. It is symmetrical about the value $x = 0$, so that the graph of P_2 is symmetrical about the vertical or *range axis* (see Figure 3).

As x gets steadily larger, the images grow by increasing steps, and therefore the graph gets steeper. But since we can consider x as large as we like, and to each x there corresponds an image, the graph can never go 'straight up'.

$\boldsymbol{-P_2(x) = -x^2}$. The values of the images for this function are numerically the same as those for the previous function for the same values of x, but the signs are now negative. Its graph is therefore as shown in Figure 4.

† 'Larger' will be used to mean numerically greater, and 'smaller' to mean numerically less. Thus, although $-3 < -1$, -3 is larger than -1 (it is a larger negative number).

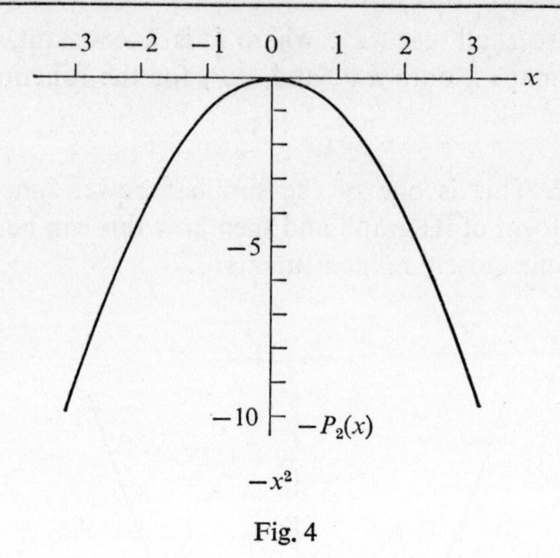

Fig. 4

$aP_2(x) = ax^2$. Functions like $3P_2$ (x maps onto $3x^2$) and $-3P_2$ (x maps onto $-3x^2$) have properties like those of P_2 and $-P_2$ respectively, but corresponding images are multiplied by 3. Figure 5 shows the relationship between the graphs of $3P_2$ and P_2.

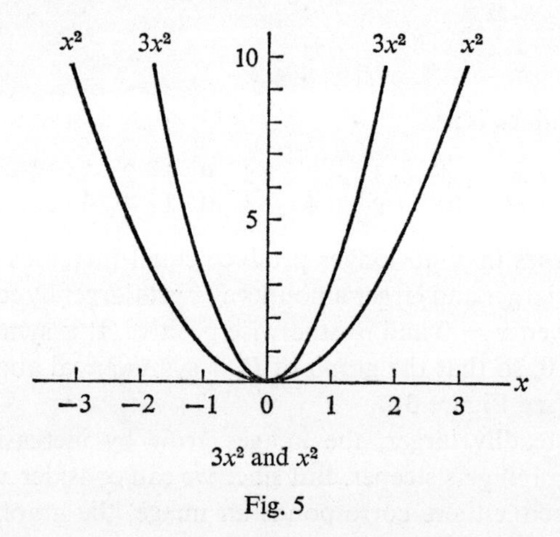

Fig. 5

$f(x) = ax^2+b$. As an example where a, b are both positive, consider

$$f(x) = 3x^2+12.$$

As $3x^2$ is never negative, $3x^2+12$ is always positive and not less than 12. The graph of $3x^2+12$ (see Figure 6) is that of $3x^2$ shifted up through 12 units, since, for corresponding values of x, $3x^2+12$ exceeds $3x^2$ by 12. $f(x)$ has a least value of 12 when $x = 0$.

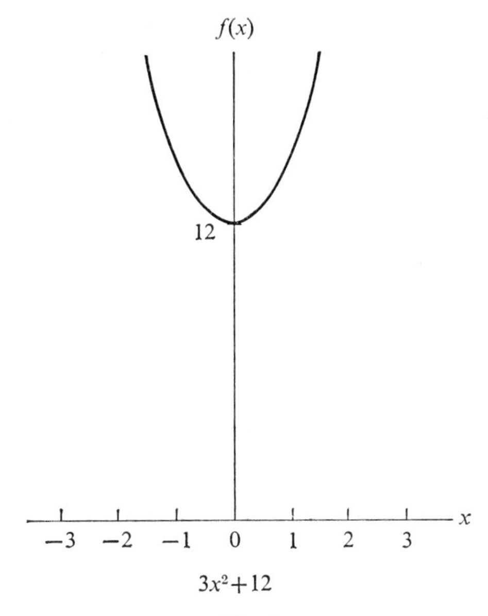

$3x^2+12$

Fig. 6

As an example where a, b have opposite signs, consider

$$f(x) = 3x^2-12.$$

Here $f(x)$ is sometimes positive and sometimes negative. Since

$$3x^2-12 = 3(x^2-4),$$

$f(x)$ is positive when $x^2 > 4$; that is, when x is either greater than 2 or less than -2. But if x lies between -2 and 2, $f(x)$ is negative. When x is 2 or -2, $f(x)$ is zero.

The least value of $f(x)$ is -12 when $x = 0$.

The graph is shown in Figure 7, and is that of $3x^2$ shifted down through 12 units.

$f(x) = (x+a)^2$. As an example where a is negative, consider

$$f(x) = (x-3)^2.$$

A table of values for $f(x)$ is:

x	-1	0	1	2	3	4	5	6	7
$(x-3)^2$	16	9	4	1	0	1	4	9	16

Compare this with the table of values for x^2, given in Section 2.1. The images are the same, but for values of x increased by 3. So the graph of f is that of P_2 shifted to the right through 3 units, as shown in Figure 8.

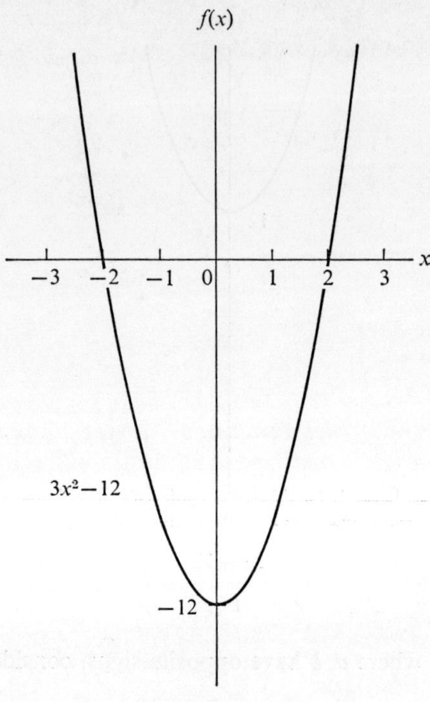

Fig. 7

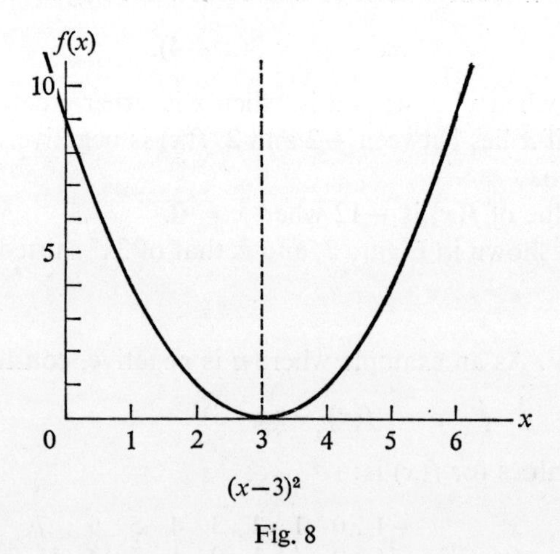

Fig. 8

The two functions therefore have the same general properties: in particular, the graph of f is symmetrical (about $x = 3$).

Exercise B

1. Sketch the graphs of:

$$3x^2+1; \quad (x-1)^2; \quad (x+2)^2; \quad -5x^2; \quad 6-2x^2.$$

2. In the same figure, sketch the graphs of $2x^2$ and $2(x-1)^2$.

3. In the same figure, sketch the graphs of $\frac{1}{2}x^2$ and $\frac{1}{2}x^2-2$.

4. Sketch the graph of $(2x-1)^2$. [*Hint:* $(2x-1)^2 = 4(x-\frac{1}{2})^2$.]

5. Sketch the graph of $(2x+3)^2$.

2.2 Higher degree power functions. Having looked at the graph of P_2 and some related functions, we shall do the same for P_3, P_4, P_5. The extension to higher power functions will be obvious. Here is a table of values from which the graphs can be sketched:

x	-4	-3	-1	-1	0	1	2	3	4
x^2	16	9	4	1	0	1	4	9	16
x^3	-64	-27	-8	-1	0	1	8	27	64
x^4	256	81	16	1	0	1	16	81	256
x^5	-1024	-243	-32	-1	0	1	32	243	1024

The graphs of P_2, P_3, P_4, P_5 are:

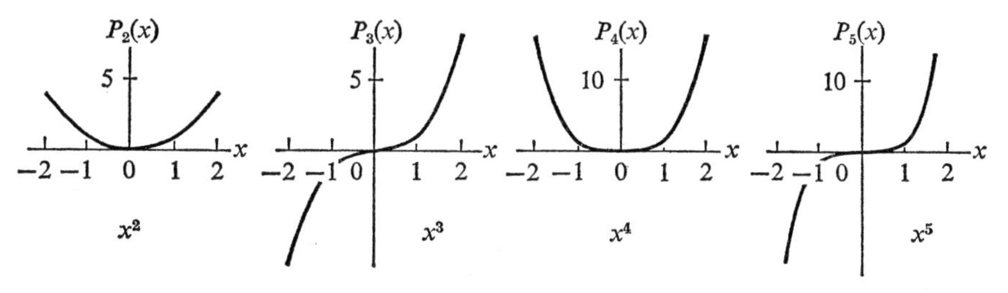

Fig. 9

Knowing the shape of the graph of x^3, it is easy to sketch the graphs of related functions of x: x^3+3 or $(x+1)^3$ for example.

Figure 10 shows the graph of x^3+3 in relation to that of x^3. It will be seen that although all points of the graph of x^3 have been shifted up by the same amount, 3 units, the curves get closer together where they are steeper.

To sketch the graph of $(x+1)^3$, as in Figure 11, all that is necessary is to shift the graph of x^3 one unit to the left.

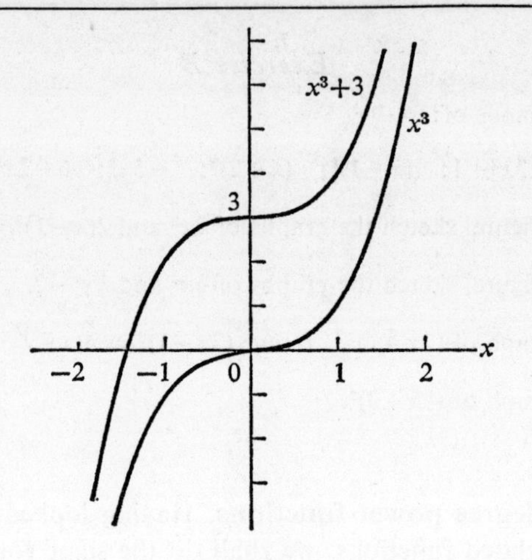

x^3+3 and x^3

Fig. 10

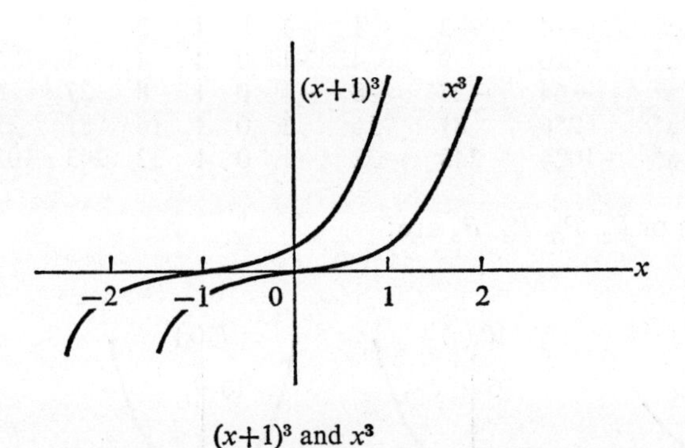

$(x+1)^3$ and x^3

Fig. 11

The justification is similar to that for the analogous problem in Section 2.1, as we see from the table of values:

x	-2	-1	0	1	2
x^3	-8	-1	0	1	8
$(x+1)^3$	-1	0	1	8	27

The reader should however accustom himself to sketching such graphs without using a table of values.

Exercise C

Sketch the graphs of:

$$-x^5;\quad 8-x^3;\quad (x-1)^3;\quad x^4+3;\quad -2(x+3)^4;\quad 5x^5-1.$$

2.3 Size of x^r. The reader will have noted the similarity between the graphs of x^2 and that of x^4; and between those of x^3 and x^5. What then are the essential differences between such functions of x?

We see this best by drawing the graphs of P_1, P_2, P_3, P_4, P_5 in the same figure. This is shown in Figure 12.

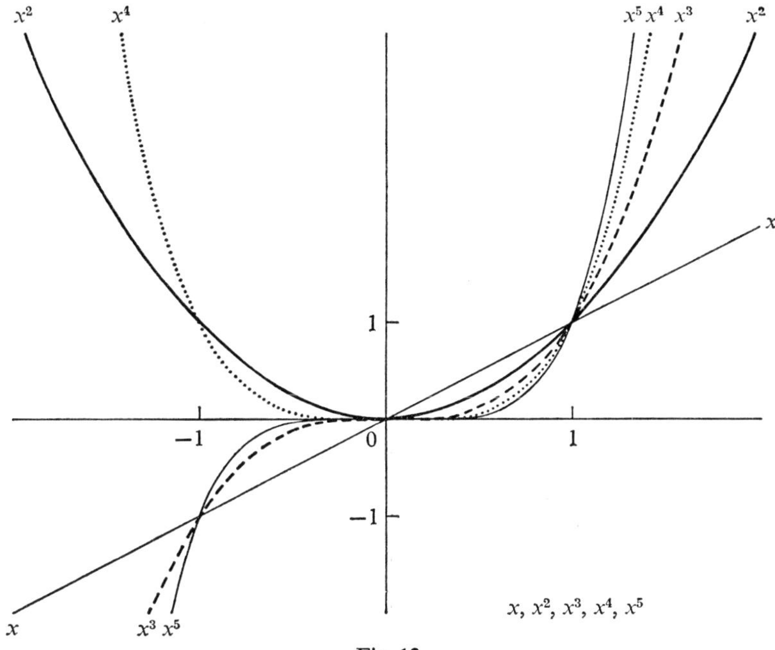

Fig. 12

The graphs of all five functions pass through (1, 1), since $P_r(1) = 1$. The graphs of odd power functions pass through $(-1, -1)$, and those of the even power functions through $(-1, 1)$.

The differences between the functions are shown in the relative numerical sizes of x, x^2, x^3, x^4, x^5.

If $x > 1$, the higher the power of x the greater the value. If x is large, clearly x^4 will be very large; but x^5 will be much larger: $10^4 = 100000$; and $10^5 = 100000$. x^5 will in fact be just x times larger than x^4.

But if x lies between 0 and 1, then the higher the power of x, the smaller the value. This can be seen in the following table:

x	0	$\pm 0{\cdot}1$	$\pm 0{\cdot}2$	$\pm 0{\cdot}4$	$\pm 0{\cdot}8$	± 1
x^2	0	0·01	0·04	0·16	0·64	1
x^3	0	$\pm 0{\cdot}001$	$\pm 0{\cdot}008$	$\pm 0{\cdot}064$	$\pm 0{\cdot}512$	± 1
x^4	0	0·0001	0·0016	0·0256	0·4096	1
x^5	0	$\pm 0{\cdot}00001$	$\pm 0{\cdot}00032$	$\pm 0{\cdot}01024$	$\pm 0{\cdot}32768$	± 1

We see that if x is smaller than 1, then x, x^2, x^3, x^4, x^5, and so on, get progressively smaller.

Again, if x is small, x^4 will be very small; but x^5 will be much smaller. If $x = 0{\cdot}01$, x^4 and x^5 are 0·00000001 and 0·0000000001—in fact x^5 is 100 times smaller than x^4.

We may sum this up by saying that, of two powers of x, for x larger than 1, the one with the higher index has the larger value; but for x smaller than 1, it has the smaller value. With slight modification this is also true of *constant multiples of powers of x*.

Take for instance $30x^5$ and $6x^4$.

It seems clear enough that for large x, $30x^5$ is the larger; not so clear that it is the smaller for small x. And indeed, what do we mean by 'small x' here?

Let us sketch their graphs. (See Figure 13.)

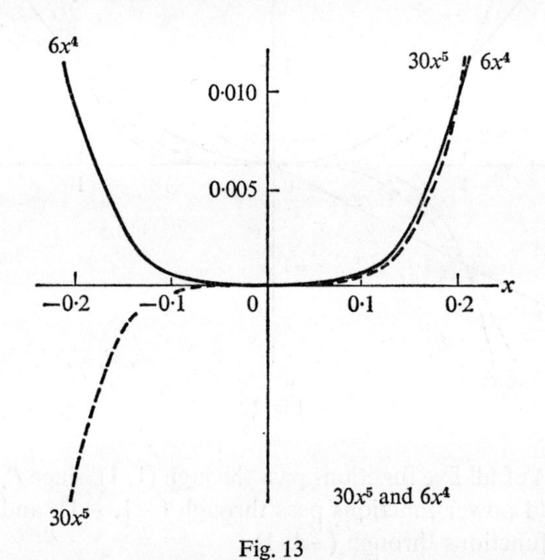

Fig. 13

Now, $$30x^5 = 6x^4$$

if $$30x^5 - 6x^4 = 6x^4(5x-1) = 0,$$

that is, if $$x = 0 \quad \text{or} \quad x = 0{\cdot}2.$$

For $0 < x < 0{\cdot}2$, $$6x^4(5x-1) < 0,$$

and hence $$30x^5 < 6x^4.$$

This justifies the graph; and we see that for *some* small x, $0 < x < 0{\cdot}2$, the higher power is the smaller. Under the definition made in the footnote on p. 23, we can include the interval $-0{\cdot}2 < x < 0$. This is covered in the restatement: the higher power has the smaller value when x is smaller than 0·2. (This will be understood, of course, to exclude $x = 0$.)

Similarly we could show, comparing their graphs, that $5000x^3 < x$ for an even smaller range of values of x bounded by $x = 0$. (The reader should do this.)

Our modification then takes the following form: *for sufficiently small x a constant multiple of the higher power will be smaller than that of the lower.*

Exercise D

Sketch the graphs of the following pairs of functions of x to show for what range of values the higher power is the larger, and for what range it is the smaller. State what the ranges are.

1. x^5, x^7. **2.** $x^6, -x^9$. **3.** $2x, \frac{1}{4}x^2$. **4.** $500x^3, 2x^4$.
5. $\frac{1}{3}x^2, 9x^5$. **6.** $-x^8, 2x^{10}$. **7.** x^{1000}, x^{1001}.

3. ADDING

3.1

Example 1. Sketch the graph of $f(x) = x^3+x-2$.

We can tackle a question like this by sketching the separate graphs of two parts of $f(x)$ (here x^3 and $x-2$) and seeing what happens when we 'add' them together. (It is convenient to speak of adding two graphs, when we mean sketching the graph of the sum of their functions.)

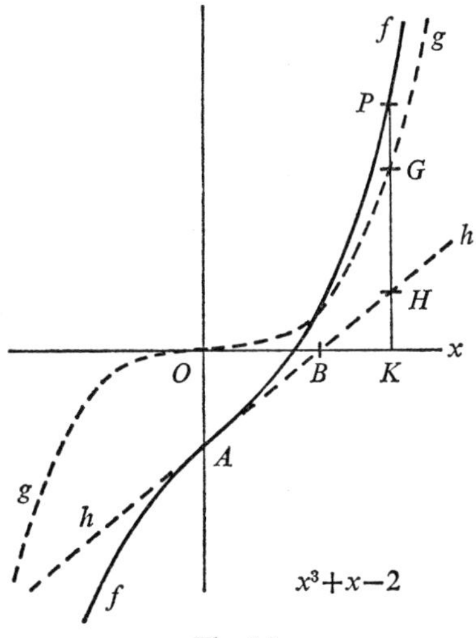

Fig. 14

The broken lines in Figure 14 show the graphs of x^3 and $x-2$. Then, at the general value of x marked by K, x^3+x-2 will be shown by a length equal to HK (representing $x-2$) plus $GK(x^3)$. Mark a point P such that $PG = HK$, and evidently P is the point we want.

The reader should now try to imagine what happens to P as K moves along the x-axis, Clearly enough as x tends to ∞, x^3+x-2 does the same, quicker than either x^3 or $x-2$.† Similarly when $x \to -\infty$.

What of the central region?

When either x^3 or $x-2$ is zero, the graph of x^3+x-2 will cross the graph of the other (see O and B). Between these, x^3 and $x-2$ are of opposite sign. Hence the graph of x^3+x-2 will lie between them, and cross the x-axis where x^3 and $x-2$ are equal and opposite in sign.

Lastly, where x is small, we are adding to $x-2$ the very small values of x^3: positive where x is positive, negative where x is negative. This shows us that here the graph of x^3+x-2 will cross that of $x-2$ *touching* it. The precise meaning of 'touching' here we shall discuss in the next chapter where this matter of adding graphs becomes critical. For the moment the reader should have no difficulty in seeing its behaviour as shown in Figure 14.

Example 2. Sketch the graph of $3x^3-x^4$.

In this example the discussion of Section 2.3 becomes critical.

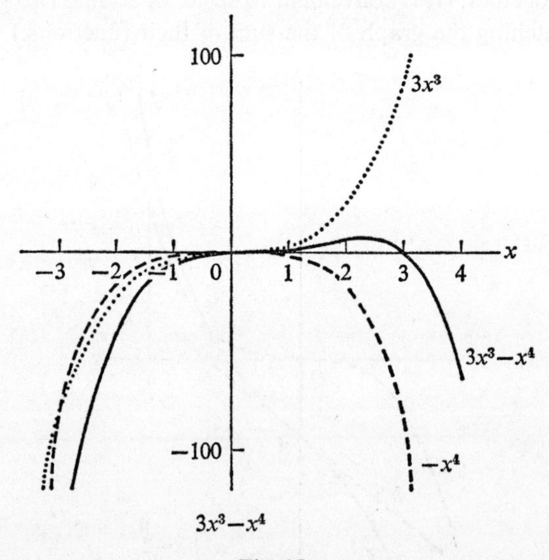

Fig. 15

† The phrase '$f(x)$ tends to ∞ as x tends to ∞' is one which in more formal analysis must be carefully defined. Here it will be sufficient for the reader to get a general impression of what is meant. The phrase 'tends to' will often be denoted by the symbol '$\to$'.

We first sketch the graphs of $3x^3$ and $-x^4$. These cross where

$$3x^3 = -x^4,$$

or
$$3x^3+x^4 = x^3(3+x) = 0,$$

or
$$x = 0 \quad \text{or} \quad -3.$$

Between 0 and -3, $-x^4$ is smaller than $3x^3$.

At $x = 3$, $3x^3$ and $-x^4$ are numerically equal but opposite in sign. Between 0 and 3, $3x^3$ is positive, $-x^4$ is negative, but here again $-x^4$ is smaller than $3x^3$.

With these points in mind, we can get the sketch shown in Figure 15. Note that near $x = 0$, the graph of $3x^3-x^4$ approaches the x-axis in the same kind of way as $3x^3$, the dominant term for small x. Note also that as $x \to \infty$, $-x^4$ dominates over $3x^3$, and gives $3x^3-x^4$ increasing negative values.

Exercise E

1. Sketch the graphs of: x^2-x+1 and x^2+x-1.

2. Sketch the graphs of: x^2+3x and $2x^2-x$.

3. Sketch the graphs of: $x+x^3$ and $x-x^3$.

4. Sketch the graphs of: x^3-3x-1 and x^3+3x+1.

5. Sketch the graphs of: x^3+2x^4; x^3-x^5; $5x^4-2x^5$.

6. Sketch the graphs of: $(x-2)^3-2x-1$; $x^3-(x+3)^2$.

4. MULTIPLYING

4.1

Example 1. Sketch the graph of $(x+1)(2x-3)$.

Here we will sketch the graphs of $x+1$ and $2x-3$ and 'multiply' them together.

The broken lines in Figure 16 show these graphs, and we note immediately that $x+1$ and $2x-3$ are zero at, respectively, -1 and $1{\cdot}5$. Clearly for these the product is itself zero; but we can see also that they naturally divide the domain into three intervals:

$$x < -1; \quad -1 < x < 1{\cdot}5; \quad x > 1{\cdot}5.$$

In the first of these we are multiplying two negative numbers, giving a positive value; in the second, a positive and a negative, giving negative; in the third two positive, giving positive.

It remains to discuss the order of magnitude of the product in these intervals. In the second, both factors remain finite, so that the product must sink to a finite minimum value between the two zeros at A and B.

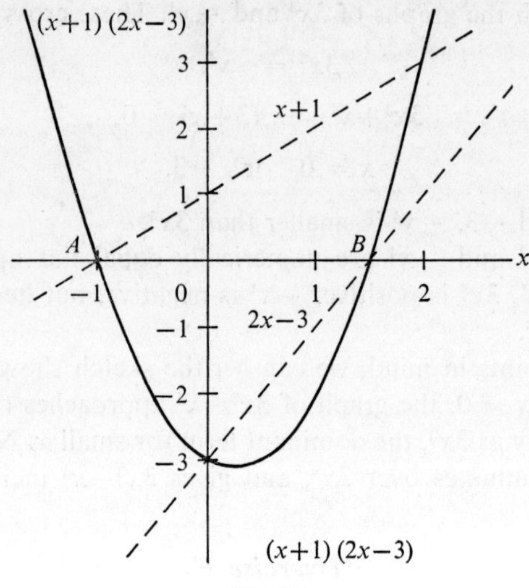

Fig. 16

In the third interval, on the other hand, both factors tend to ∞, so that the product must do the same, only faster than either.

Similarly in the first interval.

This gives the graph shown in Figure 16.

Example 2. Sketch the graph of $(x^2+1)(2-x)$.

Again, we sketch the graphs of x^2+1 and $2-x$.

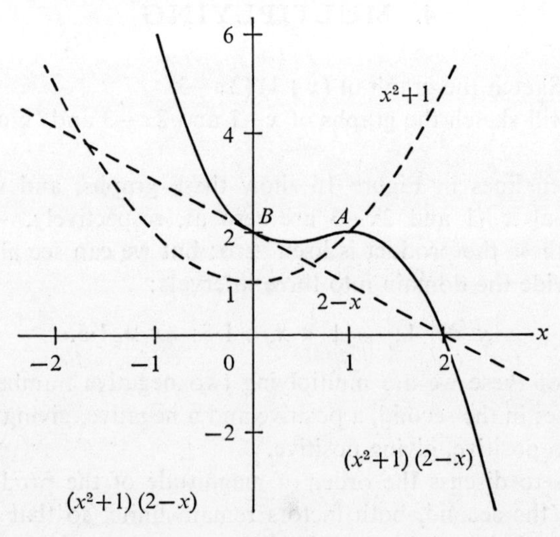

Fig. 17

Here only one factor, $2-x$, can be zero: at $x = 2$. Hence we have two intervals: $x > 2$ and $x < 2$. In the first the product is negative; in the second it is positive.

The product again tends to ∞ faster than either factor as x tends to ∞. A new point (which we did not discuss in the last example) arises in connection with those values of x for which the factors are equal to 1. Obviously if one factor is 1, the graph of the product will for that value of x cross the graph of the other factor. (We can see this also in Figure 16.) Here we see (Figure 17) that the final graph meets that of x^2+1 at A, and that of $2-x$ at B. But the latter point is exceptional. For, on either side of $x = 0$, we are multiplying $2-x$ by values of x^2+1 *just greater than* 1. (Normally, it is just greater on one side and just less on the other, as at A.) Hence the final graph, though passing through B, will on either side be just above the straight line. Hence it *touches* the graph of $2-x$ at this point.

Exercise F

1. Sketch the graphs of:

$$x(x+3); \quad (x-1)(4-x); \quad (x-3)(3x-1).$$

2. Sketch the graphs of:

$$x^2(x-2); \quad x^3(5-2x); \quad x(x+1)^3.$$

3. Sketch the graphs of:

$$x(x^2+2); \quad (x+1)(x^3+1); \quad (1-\tfrac{1}{4}x)(1-x^2),$$

noting carefully where they cross the graphs of their factors.

4. Sketch the graphs of:

$$(x-1)^2(x+1)^2; \quad x^2(x-2)^3.$$

5. RECIPROCALS

In the previous section we discussed the product of two functions. What of their quotient? We can reduce such a problem to the same form as the last if we get some sense of how the reciprocal of a function behaves. For, to take an example, $x/(x^2+1)$ can be expressed in the form $x \times \dfrac{1}{x^2+1}$ whose graph we can sketch once we get that of $\dfrac{1}{x^2+1}$.

5.1 $P_{-1}(x) = x^{-1}$. This is the simplest of the reciprocal functions. Some values are:

x	-5	-4	-3	-2	-1	0	1	2	3	4	5
x^{-1}	$-0{\cdot}2$	$-0{\cdot}25$	$-0{\cdot}33$	$-0{\cdot}5$	-1	?	1	0·5	0·33	0·25	0·2

and part of its graph, derived from these values is shown in the thick line in Figure 18.

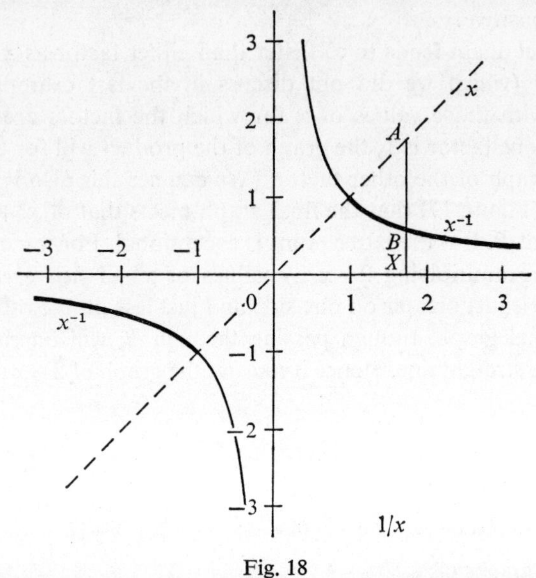

Fig. 18

But this is unsatisfactory, since we get no idea of what happens in the interval $-1 < x < 1$. 0 has no reciprocal, and has to be excluded from the function's domain. But what of values near 0? Here we see the reciprocals of

	$\frac{1}{2}$	$\frac{1}{3}$	$\frac{1}{4}$	$\frac{1}{5}$	$\frac{1}{10}$
are	2	3	4	5	10

and these give us further points on the graph. Indeed we can now see the general characteristic of the function and its graph.

For large x, the reciprocal is small (the reciprocal of 1000 is 0·001); and the larger x, the smaller its reciprocal. Hence as x tends to ∞, its reciprocal tends to zero (see footnote on p. 32) from above, although it never gets there.

Similarly if x tends to $-\infty$; though here the reciprocals are negative and tend to zero from below.

On the other hand, if x is small, the reciprocal is large (the reciprocal of 0·001 is 1000); the smaller x the larger its reciprocal. Here then as x tends to zero from above, the reciprocal tends to ∞; as it tends to zero from below, the reciprocal tends to $-\infty$.

These facts are of high importance in sketching the reciprocal of a function. The relationships can be seen by putting in the graph of x in Figure 18. Here AX represents x and BX its reciprocal. We can imagine x

tending to ∞, AX therefore becoming large and BX (its reciprocal) small; or x tending to zero, AX becoming small and BX large. Similarly if number and reciprocal are negative.

Clearly the only values for which x equals its reciprocal are those shown where the two graphs cross: $x = 1$ or -1.

5.2 Reciprocal functions.

Example 1. Sketch the graph of $\dfrac{2}{4x^2+1}$.

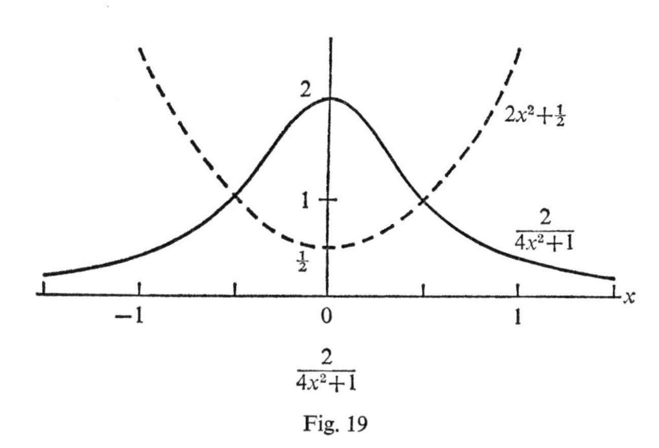

$\dfrac{2}{4x^2+1}$

Fig. 19

The reciprocal of this function of x is $\dfrac{4x^2+1}{2}$ or $2x^2+\frac{1}{2}$.

We can sketch its graph easily enough: it is that of $2x^2$ moved up by $\frac{1}{2}$.

$$2x^2+\tfrac{1}{2} = 1 \quad \text{if} \quad 2x^2 = \tfrac{1}{2};$$

that is,

$$x^2 = \tfrac{1}{4} \quad \text{or} \quad x = \pm\tfrac{1}{2}.$$

Consider now the reciprocal of this. The graph of $2x^2+\frac{1}{2}$ is shown by the broken line in Figure 19. As x tends to ∞, $2x^2+\frac{1}{2}$ becomes larger: hence the reciprocal tends to zero from above. Similarly when x tends to $-\infty$.

In the central region, the graph of the reciprocal crosses that of $2x^2+\frac{1}{2}$ where $x = \pm\frac{1}{2}$; between these, since $2x^2+\frac{1}{2}$ sinks to a minimum of $\frac{1}{2}$, the reciprocal rises to a maximum of 2.

Hence the graph is as shown in Figure 19. Since $2x^2+\frac{1}{2}$ never sinks to zero, the reciprocals never, in this example, become large.

Example 2. Sketch the graph of $\dfrac{1}{x^3+1}$.

This differs from the last example in that x^3+1 can be zero: where $x = -1$.

The graph of x^3+1 is shown in the broken line in Figure 20. The reciprocal tends to zero (from above) as x tends to ∞, and zero from below as x tends to $-\infty$.

What of the neighbourhood of $x = -1$?

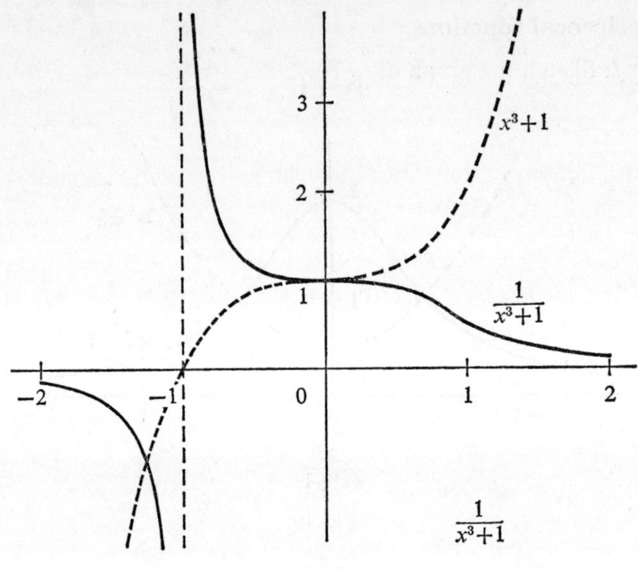

Fig. 20

As x tends to -1 from above, x^3+1 tends to zero, and its reciprocal to ∞. Similarly when x tends to -1 from below, though here x^3+1 and its reciprocal are negative.

There only remains the neighbourhood of $x = 0$, where, since the graph of x^3+1 flattens out, it seems reasonable to suppose that of its reciprocal will do the same.

Exercise G

1. Sketch graphs of $\dfrac{1}{2x+5}$ and $\dfrac{1}{1-x}$.

2. Sketch graphs of $\dfrac{1}{x^2+4}$ and $\dfrac{1}{x^2-4}$.

3. Sketch the graph of $\dfrac{1}{8-x^3}$.

4. Sketch the graph of $\dfrac{1}{x+1}$, and hence that of $\dfrac{x}{x+1}$.

5. Sketch the graph of $\dfrac{1}{x^2+1}$, and hence that of $\dfrac{x}{x^2+1}$.

6. Sketch the graphs of: $x+x^{-1}$; $2x^2+x^{-1}$.

7. Sketch the graph of x^{-2}, and that of $x+x^{-2}$.

8. Sketch in the same figure the graphs of x^{-1}, x^{-2}, x^{-3}, noting their relative sizes in various intervals of the domain.
Hence sketch the graphs of: $x^{-2}-x^{-1}$; $x^{-2}+x^{-3}$.

9. Sketch the graph of $3x^{-2}-4x^{-3}$.

10. Sketch the graph of x^2-2x (by adding) and hence that of its reciprocal.

6. FACTORS

We conclude with a method which can be applied to many of the foregoing examples, though only for those which are relatively complicated is there any advantage in using it. It depends on our ability to express a function of x as a product of linear factors; we will then be able to determine its sign in various intervals of the domain. Before going further the reader should look back to Example 1 of Section 4, where the method was, as it were, illustrated graphically.

6.1 Polynomials.

Example. Sketch the graph of $f(x) = (x+2)(x-1)(3-x)$.

Clearly the product of these three factors will only be zero when one of the factors is zero: that is, when $x = -2$, 1 or 3. It will be positive or negative according as the various factors are positive or negative.

Take a particular factor: $x+2$ say. This is zero if $x = -2$, positive if $x > -2$, negative if $x < -2$.

Let us divide the domain into four intervals separated by the three zeros -2, 1 and 3; and note the sign of each factor in each interval. This is summarized in the following table.

	$x<-2$	$-2<x<1$	$1<x<3$	$x>3$
$x+2$	$-$	$+$	$+$	$+$
$x-1$	$-$	$-$	$+$	$+$
$3-x$	$+$	$+$	$+$	$-$
$(x+2)(x-1)(3-x)$	$+$	$-$	$+$	$-$

Hence, for instance, in the interval $x < -2$, $x+2$ is negative, $x-1$ is negative, and $3-x$ is positive. It follows that the product is positive.

Likewise in $-2 < x < 1$, the product is negative; in $1 < x < 3$ it is positive; and in $x > 3$ it is negative.

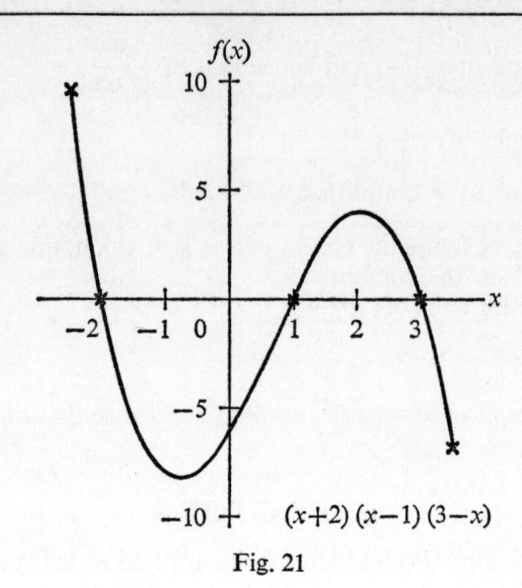

Fig. 21

If we multiply the product out, the term which dominates for large x (that is, the highest power of x) is $-x^3$. Hence as x tends to ∞, $f(x)$ tends to $-\infty$, and as x tends to $-\infty$, $f(x)$ tends to ∞.

Elsewhere the product is finite.

Combining all this information we get the sketch in Figure 21.

6.2. Product with repeated factors.

Example 1. Sketch the graph of $f(x) = (x+1)(x-2)^2$.

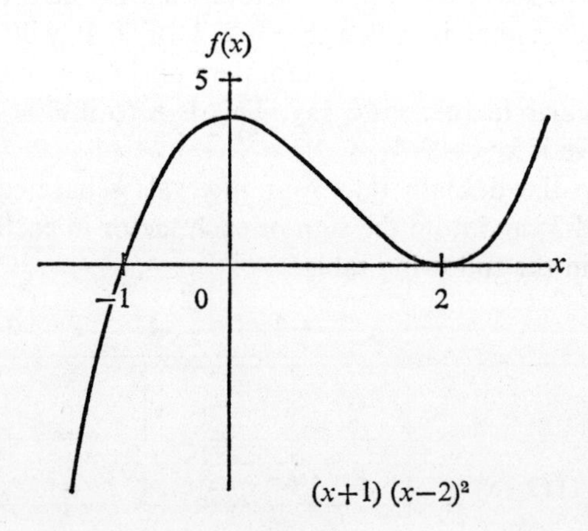

Fig. 22

A table of signs is not necessary here, since $(x-2)^2$ is always positive (or zero when $x = 2$). The sign of the product is therefore that of $x+1$, which is positive when $x > -1$ and negative when $x < -1$. When $x = -1$ or 2 the product is zero.

If the product is multiplied out, the dominant term is x^3. Hence $f(x)$ behaves like x^3 when x is large.

This is sufficient information from which to draw the graph in Figure 22.

It is interesting to compare this graph with that of the closely related function of x: $F(x) = (x+1)(x-2)(x-2{\cdot}1)$.

The sketch-graph for F is shown in Figure 23.

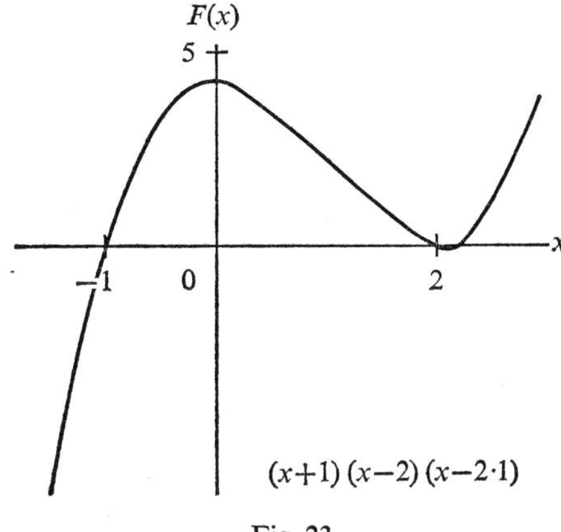

Fig. 23

The essential difference is that $F(x)$ is small and negative over the short interval 2 to 2·1. If we think of this interval shrinking to zero, so that $F(x)$ becomes $f(x)$, we can see that the graph of f will touch the x-axis at $x = 2$.

Example 2. Sketch the graph of $f(x) = (x+1)(x-2)^3$.

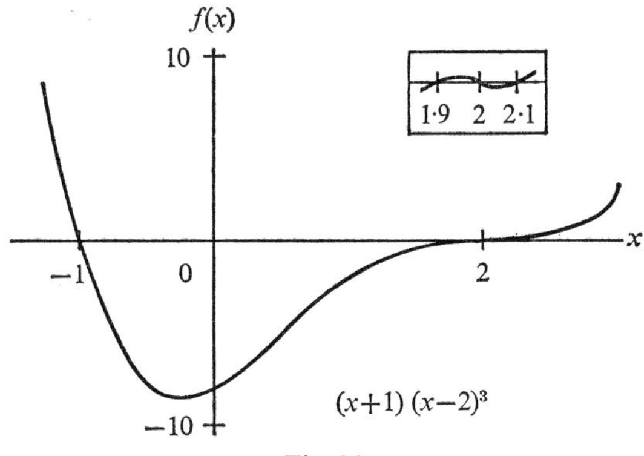

Fig. 24

Now $f(x)$ is zero when $x = -1$ or 2. It is negative when x lies between these two values and elsewhere positive. (Note that the sign of $(x-2)^3$ is the same as that of $(x-2)$.)

The dominant term for large x is x^4, so that $f(x)$ is positive and increasing as x gets larger in either direction.

For the behaviour of $f(x)$ near $x = 2$ let us consider the product $(x+1)(x-1{\cdot}9)(x-2)(x-2{\cdot}1)$ whose graph in this neighbourhood is shown in the inset to Figure 24. This graph crosses the x-axis at the points $x = 1{\cdot}9, 2, 2{\cdot}1$. What happens to this graph if we think of these three points coming together at $x = 2$? As the interval between 1·9 and 2 shrinks to zero the graph is reduced to one which touches the x-axis at $x = 2$. Similarly on the other side.

The x-axis is therefore a tangent at $x = 2$ to the graph of f; so that we have an example of a curve which crosses its own tangent at its point of contact. We have already met this, of course, with the graph of x^3. Indeed it will happen whenever there is a triple (or any odd multiple) repeated factor.

6.3 Rational functions.

Rational functions of x are the ratios of polynomials. If the polynomial denominator factorizes into linear factors, $f(x)$ will tend to ∞ as x tends to each value which makes a factor zero. The next two examples illustrate this feature of such rational functions.

Example 1. Sketch the graph of $f(x) = \dfrac{1}{(x+1)(x-3)}$.

It can readily be shown that $f(x)$ is positive if $x < -1$ or $x > 3$. Between -1 and 3, $f(x)$ is negative.

What happens near $x = 3$? Here $x+1$ is approximately 4 while $x-3$ is small; and therefore $f(x)$ is large. As x tends to 3 from above, $f(x)$ tends to ∞. And as x tends to 3 from below, $f(x)$ tends to $-\infty$. In the same sort of way, $f(x)$ tends to ∞ or $-\infty$ as x tends to -1.

When x is large, the denominator is large and positive; hence, $f(x)$ is small and positive. As x tends to ∞, $f(x)$ tends to zero from above.

From this information, and the fact that $f(x)$ is never zero, the graph shown in Figure 25 can be drawn.

An alternative way of sketching the graph is to consider the reciprocal of $(x+1)(x-3)$. Its graph is shown dotted in the figure.

Example 2. Sketch the graph of $f(x) = \dfrac{x-1}{(x+1)(x-3)}$.

The sign of $f(x)$ is the same as that of $(x-1)(x+1)(x-3)$. It is shown in the table at the top of p. 44.

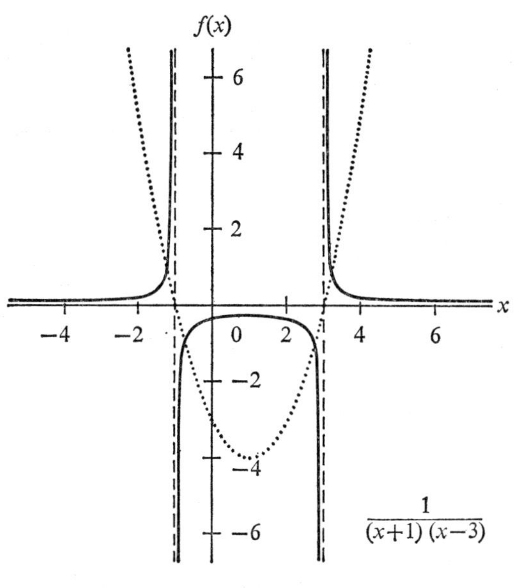

$$\frac{1}{(x+1)(x-3)}$$

Fig. 25

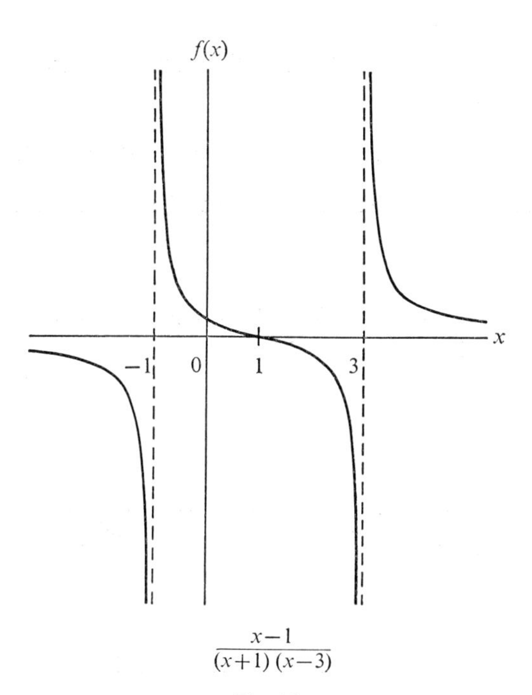

$$\frac{x-1}{(x+1)(x-3)}$$

Fig. 26

		-1	1	3
$x-1$	$-$	$-$	$+$	$+$
$x+1$	$-$	$+$	$+$	$+$
$x-3$	$-$	$-$	$-$	$+$
$f(x)$	$-$	$+$	$-$	$+$

Consider values of x near to -1. For such values, $x-1$ and $x-3$ are approximately -2 and -4, while $x+1$ is small. Hence $f(x)$ is large; and tends to $\pm\infty$ as x tends to -1 from above or below. Similarly near $x = 3$.

Next consider large values of x. The dominant term in the numerator is x, and in the denominator x^2. So for large x, $f(x)$ behaves like $1/x$.

We can now sketch the graph, shown in Figure 26, after noting that $f(x)$ is zero *only* when $x = 1$.

Exercise H

1. Sketch the graphs of:

$x(x-2)$; $(3-x)(x+1)$; $x(x-3)(2x-1)$; $(3x+2)(x-5)(4-x)$; $x(x-1)(x-2)(x-3)$.

2. Sketch the graphs of:

$$x^2(2x-1);\quad x^3(x+2);\quad x(3-x)^2;\quad x(3+x)^3.$$

3. Sketch the graphs of:

$$(x+2)(3-x)^2;\quad (x+2)^2(3-x)^2;\quad (x-2)^3(3-x)^2.$$

4. Find functions with the general character of the graphs shown in the figure:

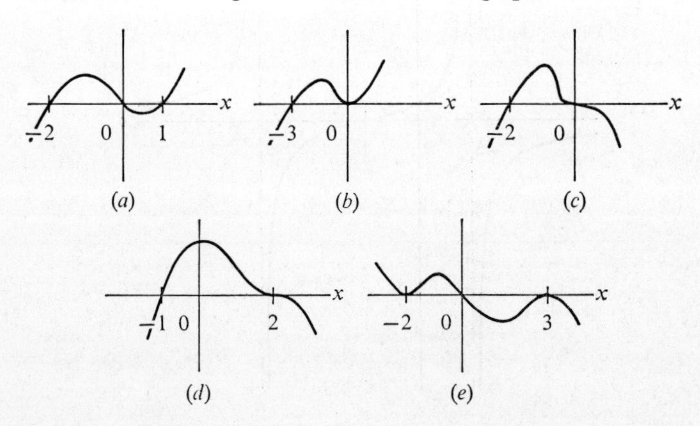

5. Factorize the following functions of x, and hence sketch their graphs:

$$x^2-3x;\quad x^3+2x^2;\quad x-x^3;\quad 4x^5+3x^4;\quad x^3+2x^4.$$

6. Make a table of signs for the following functions of x, and use it to help sketch their graphs:

$$\frac{1}{(x-2)(x-3)};\quad \frac{1}{(x+1)(x-2)^2};\quad \frac{1}{x^2-9};\quad \frac{x-2}{x(x-3)};\quad \frac{x^2+2x}{x-1}.$$

7. Sketch graphs of: $\quad \dfrac{x}{2x+1};\ \dfrac{x^2}{2x+1};\ \dfrac{x^3}{2x+1}.$

Miscellaneous Exercise

1. When a firm manufactures n cars each day the cost of production of each car is £C where $C = an+\dfrac{b}{n}$, a and b being constants. Given that $C = 800$ when $n = 100$ and $C = 700$ when $n = 200$, determine the constants a and b. By drawing the graph of C as a function of n, for values of n from 100 to 200, find the least cost of production of a car, and the number of cars per day corresponding to this least cost. Could there be any other value of n outside the interval 100 to 200 for which C could be less than its 'least' value within this interval?

2. An object is placed to the left of a fixed lens of focal length 2 m and the image is formed either to the right or left of the lens depending on the position of the object. If the object and image are u metre and v metre from the lens, where u is negative and v is positive or negative according as the image is to the right or left of the lens, u, v are related by

$$\frac{1}{2} = \frac{1}{v} - \frac{1}{u}.$$

Express v as a function of u, and plot the graph of the function. Where must the object be placed if the image is to be less than 4 m from the lens? Where can images not be formed?

3. A man rows downstream for a distance 200 m, at a speed of 5 m/s relative to the water. He then turns round and rows back again at 10 m/s relative to the water. Express the time he takes as a function of the speed of the water. Sketch the graph of this function, and hence find the least time in which he can complete the course.

4. In establishing an air-speed record, a pilot must fly over a course of given length, first directly against the wind, and then back again with it, his speed being judged as his average speed over the ground. If the speed of the wind is 40 km/h, express his average ground-speed as a function of his speed relative to the air (assumed constant). How fast must he go if his average ground-speed is to differ from his air-speed by less than 1 km/h?

5. A hollow cone of height h cm fits exactly round a sphere of radius 1 cm. Show that its volume is $\pi h^2/(h-2)$ cm³. Sketch the graph of this function of h, and hence estimate roughly the volume of the smallest cone that can fit round the sphere in this way.

3

THE DERIVED FUNCTION

In the last chapter we examined roughly the graphs of some simple functions. Here we shall look, in more detail, at their behaviour near particular points. Of critical importance in this discussion is the question 'How small?', to which we shall attend in the first section.

1. SIZE IN A NEIGHBOURHOOD

1.1 A Project. It is hoped to examine the size of such functions of x as $\frac{1}{4}x$, $3x^2$, $200x^3$ near $x = 0$.

$x =$ 0·01	0·02	0·03	0·04	0·05	0·2	0·4	0·6	0·8
$x^2 =$ 0·0001	0·0004	0·0009	0·0016	0·0025	0·04	0·16	0·36	0·64
$x^3 =$ 0·000001	0·000008	0·000027	0·000064	0·000125	0·008	0·064	0·216	0·512

Using the table of values, draw accurate graphs of the following groups of functions of x, using the same scale and axes for each group.

(*a*) $2x$, $3x^2$, $5x^3$ for $x = \pm 0{\cdot}1, \pm 0{\cdot}2, \pm 0{\cdot}4, \pm 0{\cdot}6, \pm 0{\cdot}8$.

(*b*) $3x^2$, $\frac{1}{100}x$ for $x = \pm 0{\cdot}01, \pm 0{\cdot}02, \pm 0{\cdot}03, \pm 0{\cdot}04, \pm 0{\cdot}05$.

(*c*) $3x^2$, $\frac{1}{100}x$ for $x = \pm 0{\cdot}001, \pm 0{\cdot}002, \pm 0{\cdot}003, \pm 0{\cdot}004, \pm 0{\cdot}005$.

(*d*) $2000x^3$, $\frac{1}{4}x$ for some range of values of x which will show that near enough to $x = 0$, $2000x^3$ becomes less than $\frac{1}{4}x$, but that it is not always less than or equal to $\frac{1}{4}x$.

(*e*) $2x$, $2x+3x^2$ for the same values of x as in (*a*).

(*f*) $1-2x$, $1-2x+5x^3$ for the same range of values as in (*a*).

Comparing your results, can you make any generalizations about the relative size of such functions as x, $3x^2$, $5x^3$ near $x = 0$ and far from $x = 0$? Or about the behaviour of such functions as $\frac{1}{4}x-3x^2$ or $\frac{3}{4}+\frac{1}{4}x+20x^3$ near $x = 0$?

Keep your work for discussion in Sections 1.2 and 1.3.

1.2 Small compared with x. Let us turn now to the question 'How small?' For instance, near $x = 0$, is $3x^2$ smaller than $2x$? Or $3x^2$ than $\frac{1}{100}x$? Or is $5x^3$ small compared with x? (By smaller or bigger in this chapter, we shall mean numerically smaller or bigger, without regard to sign. Thus $-0{\cdot}3$ is bigger than $0{\cdot}2$.)

The answer to these questions depends, obviously, on what we mean by 'near $x = 0$'. Put $x = 0{\cdot}01$ (which seems near 0) and we find that $3x^2$ is bigger than $\frac{1}{100}x$. (See graph (*b*).) On the other hand, put $x = 0{\cdot}002$ (which is nearer) and we find that, not only is $3x^2$ smaller than $\frac{1}{100}x$, *but that for all non-zero x between* 0·002 *and* −0·002 *it is smaller than* $\frac{1}{100}x$.

Now, in general, by any statement of the form

'$3x^2$ is smaller than $\frac{1}{100}x$ near $x = 0$'

or

'$2000x^3$ is smaller than $\frac{1}{4}x$ near $x = 0$'

we mean that *there exists an interval containing* $x = 0$ *within which (except at* $x = 0$ *itself) this is always true*. We exclude $x = 0$ since there $3x^2 = \frac{1}{100}x$ and $2000x^3 = \frac{1}{4}x$.

Let us now examine the two statements in the light of this definition.

Is $3x^2$ smaller than $\frac{1}{100}x$ near $x = 0$? Graph (*b*) does not tell us. (See Figure 1.) There the range of values chosen was too wide to offer convincing evidence. But graph (*c*) does. (See Figure 1.) There we see that the interval within which $3x^2$ is smaller than $\frac{1}{100}x$, except at $x = 0$, stretches approximately from

$$x = -0{\cdot}0033 \quad \text{to} \quad x = 0{\cdot}0033.$$

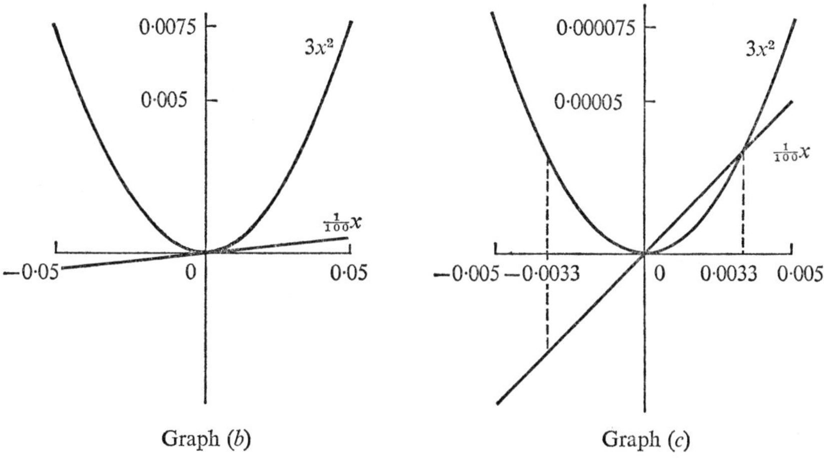

Fig. 1

Is $2000x^3$ smaller than $\frac{1}{4}x$ near $x = 0$? Suppose that for (*d*) we choose the range of values

$$x = \pm 0{\cdot}01, \quad \pm 0{\cdot}02, \quad \pm 0{\cdot}03.$$

Figure 2 shows the result of this. We find an interval, stretching approximately from $x = -0{\cdot}012$ to $x = 0{\cdot}012$, within which, except at $x = 0$, $2000x^3$ is smaller than $\frac{1}{4}x$.

The statement is, therefore, true.

Now these results are capable of extension. We have seen that not only is

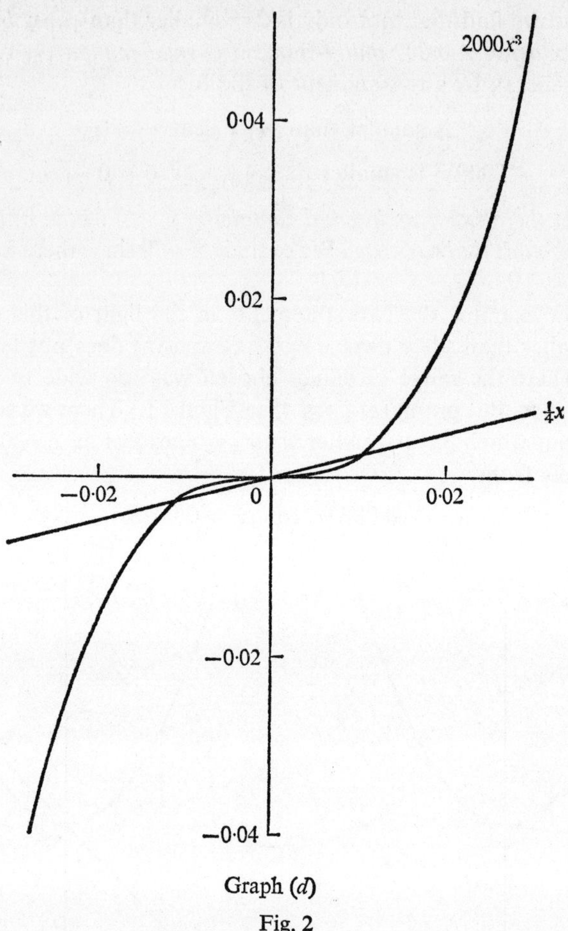

Graph (*d*)

Fig. 2

$3x^2$ smaller than $2x$ near $x = 0$ (see graph (*a*)); it is also smaller than $\frac{1}{100}x$. Likewise we could show that, near $x = 0$, it is smaller than $\frac{1}{1000}x$, than $\frac{1}{10000}x$, and so on. Indeed we could show that, near $x = 0$, $3x^2$ is smaller than anything of the form kx, no matter how small k may be provided it is not zero.

Take for example $\frac{1}{50000}x$.

$$3x^2 = \frac{1}{50000}x \quad \text{if } x = 0 \quad \text{or} \quad \text{if } x = \frac{1}{150000} \approx 7 \times 10^{-6}.$$

If we draw the graphs of $3x^2$ and $\frac{1}{50000}x$ from $x = -8 \times 10^{-6}$ to $x = 8 \times 10^{-6}$ we shall find an interval within which, except at $x = 0$, $3x^2$ is the smaller.

Of course the smaller k, the smaller the corresponding interval will be; but the interval will always be there. And it is for this reason that—as the reader should have noted for himself—the graphs of $3x^2$, $5x^3$, $2000x^3$ and functions of x like them, *flatten out* near the origin and cling close to the x-axis, sidling up to it as it were, as no graph of a linear function can do.

For consider graphs (*a*) and (*c*). They show us that in some interval containing the origin, $3x^2$ is smaller than $2x$.† In a smaller interval it will be smaller than x; in another smaller again, than $\frac{1}{2}x$; than $\frac{1}{5}x$; than $\frac{1}{10}x$; in yet another (as graph (*c*) shows us), than $\frac{1}{100}x$; and so on, indefinitely. Now the graphs of these functions of x are the radial spokes we show, dotted, in Figure 3, each with a smaller gradient than the last. We could go on drawing a sequence of such lines for ever, each more nearly horizontal than the last, each nearer the x-axis. And the graph of $3x^2$, for no matter how small an interval, would have to squeeze under each one of them.

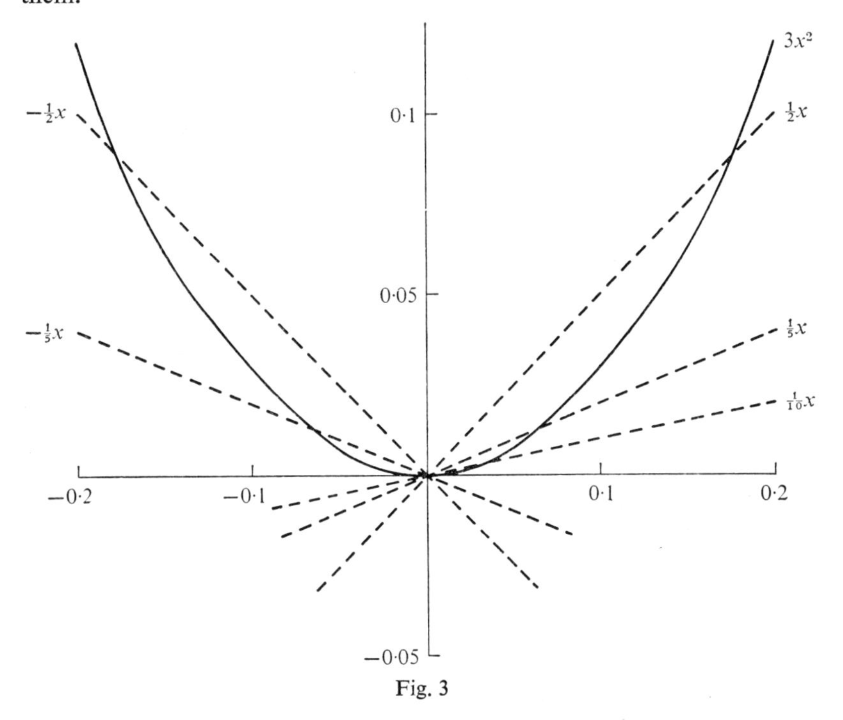

Fig. 3

That is what we mean by saying that the graph of $3x^2$ *flattens out* near the origin as does that of no linear function. We can put it geometrically by saying that the graph of $3x^2$ *touches* the x-axis at the origin.

Of course, in the instance we have chosen, the graph of $3x^2$ only squeezes under the others where x is positive, because only there have $3x^2$ and $2x$, x, $\frac{1}{2}x$, ... the same sign. One might say then that we have only showed that it touches to the *right* of the origin.

† Except, again, at $x = 0$ itself. We shall leave out this stipulation in future.

This however is not so. For the interval within which $3x^2$ is smaller than any of the others has to include the origin. Therefore on the other side too $3x^2$ is—for however small an interval—numerically the smaller.

We can see this graphically by also putting in Figure 3 the reflections of the dotted lines in the x-axis. To the left of the origin, the graph of $3x^2$ has also to squeeze under these.

1.3 Tangency. This notion of touching, or *tangency*, leads to another important point. It should be clear from graphs (*e*) and (*f*).

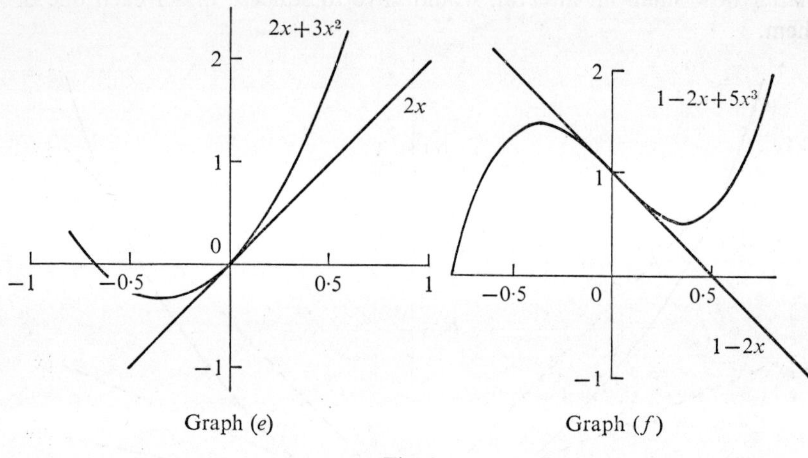

Fig. 4

If the reader looks at these (in Figure 4), he will see that the graphs of $2x+3x^2$ and $1-2x+5x^3$ touch those of $2x$ and $1-2x$. And in general we might guess that, wherever we add something like $3x^2$ or $5x^3$ to a linear function of x (here $2x$ and $1-2x$ respectively) the graph of the sum touches that of the linear function.

But what do we mean by 'touch' here?

Let us amplify this by considering an independent example:

$$f(x) = 1+\tfrac{1}{10}x+\tfrac{1}{2}x^3.$$

See Figure 5.

We can show that for some interval containing the origin, $\frac{1}{2}x^3$ is smaller than $\frac{1}{10}x$. Likewise for $\frac{1}{20}x$, for $\frac{1}{50}x$, for $\frac{1}{100}x$, and so on.

It follows that there is an interval of positive x, bounded by the origin, in which $f(x)$ lies between $1+\frac{1}{10}x$ and $1+\frac{1}{10}x+\frac{1}{10}x$, so that in this interval its graph must lie between those of the two linear functions of x.

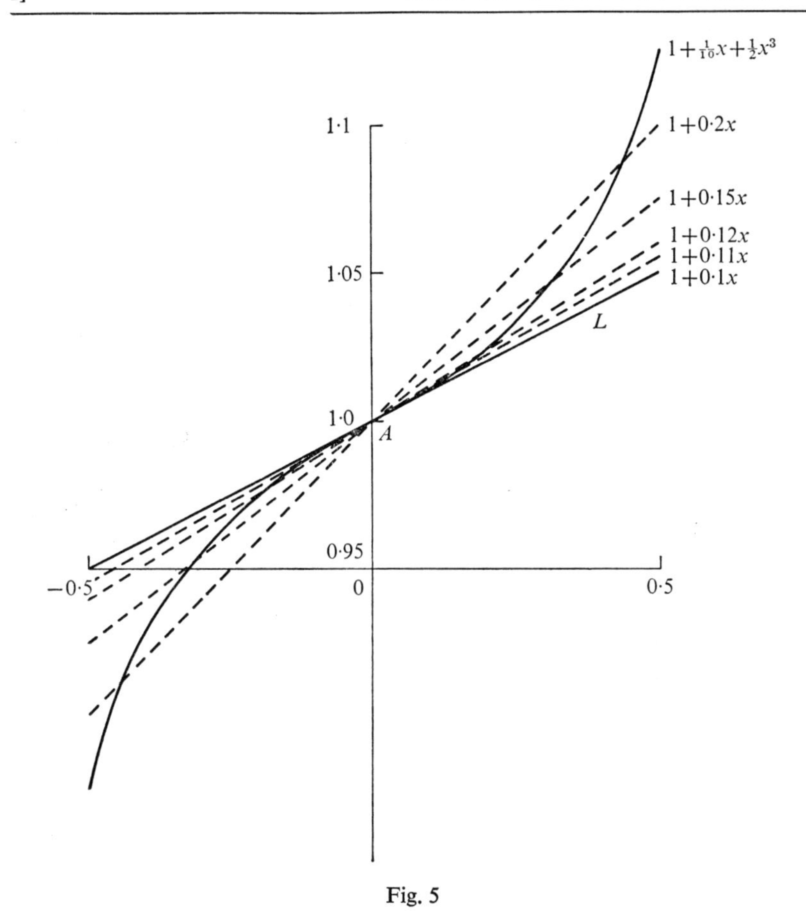

Fig. 5

Likewise its graph, for a similar interval, must lie between those of $1+\frac{1}{10}x$ and $1+\frac{1}{10}x+\frac{1}{20}x$. And so on.

Now the graphs of

$$1+\tfrac{1}{10}x+\tfrac{1}{10}x,\quad 1+\tfrac{1}{10}x+\tfrac{1}{20}x,\quad 1+\tfrac{1}{10}x+\tfrac{1}{50}x,\quad 1+\tfrac{1}{10}x+\tfrac{1}{100}x,\quad \ldots$$

or of $$1+0{\cdot}2x,\quad 1+0{\cdot}15x,\quad 1+0{\cdot}12x,\quad 1+0{\cdot}11x,\quad \ldots$$

all pass through the point (0, 1), which we have marked A. They are straight lines with gradients 0·2, 0·15, 0·12, 0·11, ... In fact they are the dotted radial spokes shown in Figure 5: a sequence getting closer and closer to the graph of $1+2x$ (marked L).

But the graph of $f(x)$ has to squeeze between each one of them and L, for no matter how small an interval. Therefore we see it again sidling up to L, clinging to it close as did the graph of $3x^2$ to the x-axis in Figure 3.

Now this is what we mean by saying that the graphs of $f(x)$ and $1+0{\cdot}1x$ touch at A, or that the graph of $1+0{\cdot}1x$ is a tangent to that of $f(x)$:

> *that if we draw a line through A, no matter how close to L, the graph of f(x) has to lie between L and the line (or its reflection in L) for some interval bounded by A.*†

1.4 Small compared with x. All in all, the distinction between functions of x like $2x$, $\frac{1}{10}x$ on the one hand, and $3x^2$, $5x^3$ on the other is so important that we introduce a phrase to distinguish the latter.

We say they are *small compared with x near $x = 0$*, or simply *small compared with x*. In full:

> *a function of x, $f(x)$, is small compared with x near $x = 0$ if, for each positive k, we can find an interval containing $x = 0$ within which, except at $x = 0$ itself, $f(x)$ is numerically less than kx.*

This is, as a matter of fact, quite a formal definition. That said, however, the reader need not at this stage bother himself very much about its rigour. It is sufficient to grasp the idea.

We introduce also a notation to denote functions of x small compared with x. This is s_x.

Any function of x small compared with x can simply be written s_x.

'$F(x) = s_x$' will be read: '$F(x)$ is small compared with x'.

Something like '$1-3x+s_x$' will be read: '$1-3x$ plus a function of x small compared with x'.

Not only squares and cubes but all higher powers of x and their multiples are small compared with x: for instance $4x^5$, $-x^{2{\cdot}5}$, $3x^{22}$. (See Exercise A, Question 1.)

There are various formal laws. For instance, the sum of two functions small compared with x is itself small compared with x. (See Exercise A, Question 12.) For further laws, which will be useful later in this book, see Exercise A, Questions 10, 11, 13.

Much of this however can be left until the future. The essential thing to grasp at the moment is the geometrical property we have already discussed: that the graph of $2-3x+s_x$ (for instance) *touches* that of $2-3x$ at (0, 2). The property will be further illustrated in the following example.

Example. Sketch roughly the graphs of

$$3-2x-x^2, \quad 3-2x+2x^2, \quad 3-2x+5x^3 \quad \text{near} \quad x = 0.$$

Each of these functions of x is of the form $3-2x+s_x$, and we may therefore say that their graphs all touch that of $3-2x$ at (0, 3). However

† Again, of course, we have only shown the curve touches L to the right of A. But it is not hard to adapt the data in order to show it touches at the left as well.

we cannot say in what way they touch without another glance at the actual nature of the terms small compared with x.

See Figure 6. The dotted lines passing through the origin show the rough graphs of $-x^2$, $2x^2$ and $5x^3$ near $x = 0$. If we use the techniques for 'adding' graphs which we developed in Chapter 2, it is easy enough to see that for $3-2x-x^2$ we get a graph touching the tangent from beneath; for $3-2x+2x^2$ a graph touching it from on top; and for $3-2x+5x^3$ a graph touching from above on the right and from below on the left: that is, crossing the tangent.

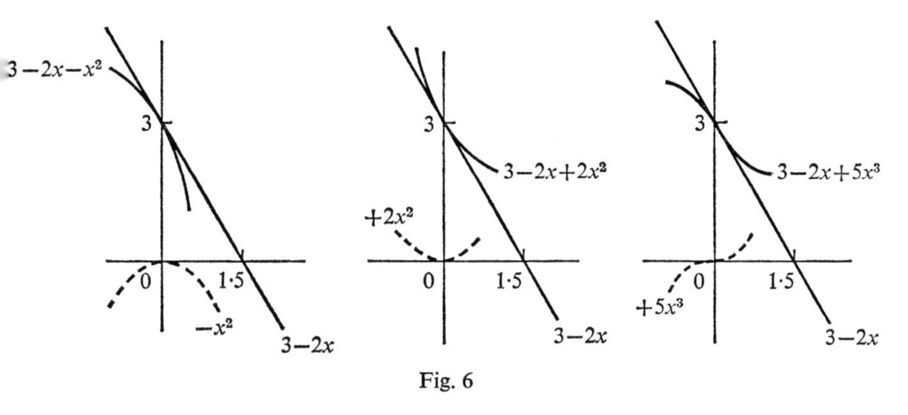

Fig. 6

This last manner of touching is, of course, exceptional, and can only happen at isolated points on a smooth curve.

Exercise A

1. Show, by drawing their graphs from $x = 0$ to $x = 1$, that there is an interval in which $3x^4$ is smaller than $2x$. Show there is an interval in which it is smaller than x, and one in which it is smaller than $\frac{1}{4}x$. Indicate the rough bounds of these intervals.

Draw a new graph with a range of your own choice to show clearly there is an interval in which $3x^4$ is less than $\frac{1}{100}x$. What is the interval?

2. Which of the following is small compared with x?

$$3x^4,\quad 4x^3,\quad 478x^2,\quad 9000x,\quad \tfrac{1}{9000}x,\quad x^{92},\quad x^{-1},\quad x^2-10^{-6},\quad x^2 10^{-6}.$$

3. With the aid of an accurately drawn figure, show that the graph of $2x+\frac{1}{40}x$ does not touch that of $2x$.

$\frac{1}{40}x$ is always smaller than $2x$, than x, than $0{\cdot}1x$. What essential characteristic does it lack that prevents it from being small compared with x?

4. Draw accurate graphs of the following functions of x near $x = 0$, and find out whether they are small compared with x: $\sqrt{x}$, $x\sqrt{x}$.

5. By drawing accurately the graph of $x^2+0\cdot0001$ near the origin (from $x = -0\cdot1$ to $x = 0\cdot1$) find whether there is an interval in which it is smaller than $0\cdot04x$. Repeat for $0\cdot03x$.

Are there values of k for which $x^2+0\cdot0001$ is *never* less than kx? If so, give an example of one such value, and say what you can about them.

Is $x^2+0\cdot0001$ small compared with x?

6. *Sketch* the graphs of the following, and from your sketches decide whether they are small compared with x:

$$\frac{x^2}{x-2}, \quad \frac{x^2}{x-0\cdot01}.$$

7. Sketch roughly the graphs of the following near $x = 0$:

$$3+2x, \quad 3+2x+2x^2, \quad 3+2x-\tfrac{1}{2}x^2, \quad 3+2x-\frac{x}{100}, \quad 3+2x+2x^3, \quad 3+2x-x^6.$$

8. Sketch roughly the graphs of the following near $x = 0$:

$$3-\tfrac{1}{2}x-x^2, \quad 3-\tfrac{1}{2}x+2x^3, \quad 3-\tfrac{1}{2}x-x^5.$$

9. Sketch roughly the graphs of the following near $x = 0$:

$$2x-3x^3, \quad 2-x^2, \quad -5+x^2, \quad 3+x^3, \quad 3-10^{-6}x.$$

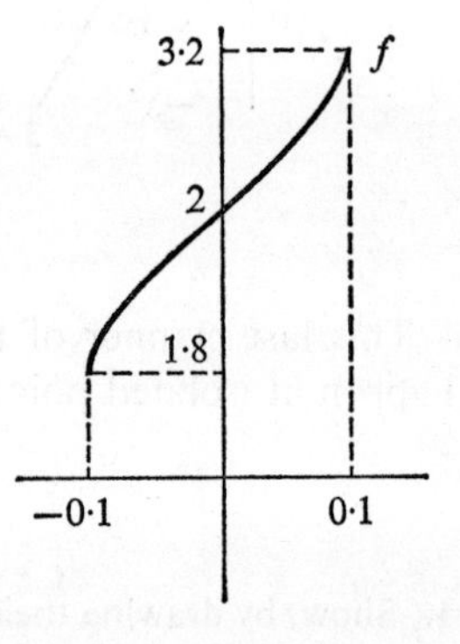

10*. $f(x)$ is a function of x whose graph near $x = 0$ is shown in the figure. Discuss whether $xf(x)$, $x^2f(x)$ and $x^3f(x)$ are small compared with x.

Can you suggest a function f (of some type which you have already met) for which $x^2f(x)$ is *not* small compared with x?

Can you suggest a condition on f for $x^2f(x)$ to be small compared with x?

11*. The graph of $f(x)$ is shown roughly in the figure. Discuss whether $xf(x)$ is small compared with x.

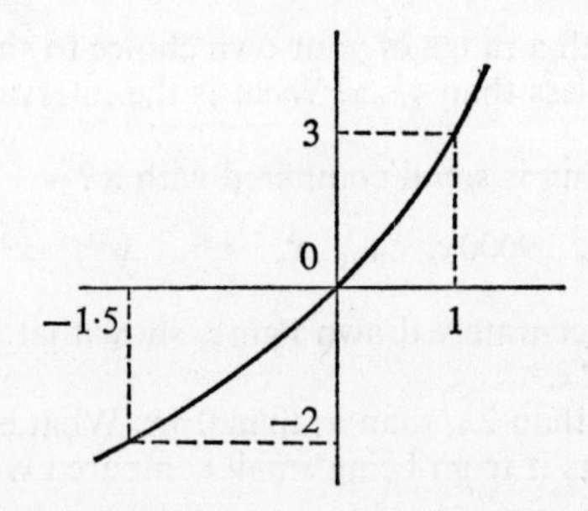

Can you suggest a sufficient condition on g for $xg(x)$ to be small compared with x?

12*. $f(x)$ and $g(x)$ are both small compared with x. By considering the intervals in which they are smaller than $\frac{1}{20}x$, prove there is an interval containing the origin within which $f(x)+g(x)$ is smaller than $\frac{1}{10}x$.

Show that $f(x)+g(x) = s_x$.

13*. Show that if $f(x) = s_x$, then $xf(x) = s_x$.

14*. If for each non-zero k an interval can be found containing the origin within which (except at the origin itself) $f(x)$ is smaller than kx^2, then $f(x)$ is said to be small compared with x^2.

Draw graphs, or use calculation to show that $10x^3$ is small compared with x^2.

Sketch the graphs of x^2 and x^2-10x^3 to show clearly how they lie in relation to each other near the origin.

15*. Sketch roughly the graphs of the following three functions of x all in the same figure, showing clearly where that of the last cuts that of the first:

$$1-\tfrac{1}{2}x, \quad 1-\tfrac{1}{2}x-x^2, \quad 1-\tfrac{1}{2}x-x^2+\tfrac{1}{4}x^3.$$

16*. Repeat Question 15 for the following functions of x:

$$-3x, \quad -3x+2x^3, \quad -3x+2x^3-x^4.$$

2. FORM OF GRAPH NEAR GENERAL POINT

2.1 General point. So far we have considered the form of a function's graph where it cuts the vertical axis. Can we now extend this method to deal with any point on its graph? Can we, for instance, find the form of the graph of a function f near $x = 2$? (See Figure 7.)

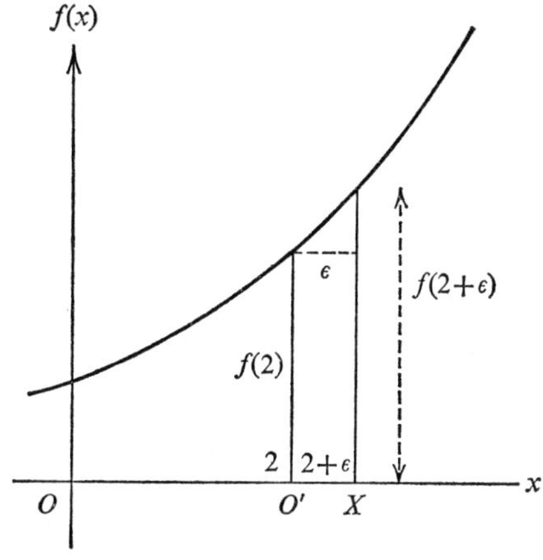

Fig. 7

Now to follow the same kind of argument, we must express $f(x)$ in terms of a variable which is small near $x = 2$. The simplest to choose would be $x-2$, represented in Figure 7 by the length $O'X$. Let us call it ϵ. Then OX is $2+\epsilon$, and $f(x)$ can be written $f(2+\epsilon)$.

ϵ, note, is to be thought of as a variable supplanting x. It can take positive and negative values, and by considering small values of ϵ we can explore the form of the graph of f near $x = 2$ (or near $\epsilon = 0$), provided of course we can express $f(x) = f(2+\epsilon)$ conveniently in terms of ϵ.

Methods for doing this we shall study later. But suppose, for the moment, that we *know*

$$f(2+\epsilon) = 3+2\epsilon-\epsilon^2.$$

Now the graph of $3+2\epsilon$ is a straight line.

When $\epsilon = 0$, we get $f(2) = 3$; and whenever we increase ϵ by α (say), we increase $3+2\epsilon$ by 2α. Thus we get as graph the line passing through (2, 3) with gradient 2.

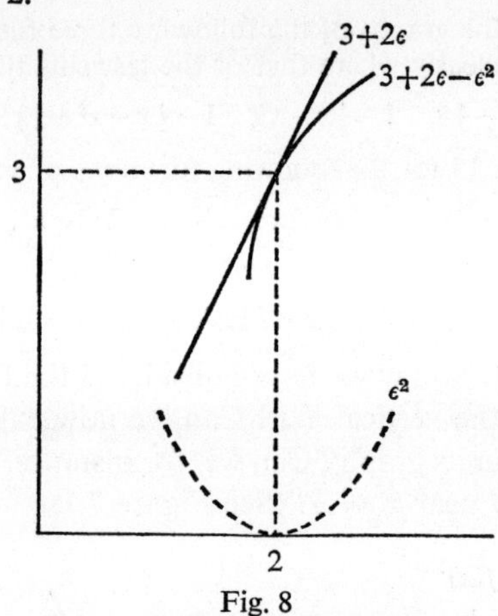

Fig. 8

We now find the graph of $3+2\epsilon-\epsilon^2$ by subtracting that of ϵ^2—a function of ϵ small compared with ϵ—whose graph is shown by the dotted line in Figure 8. Hence an argument similar to that in the last section shows that the graph of f touches that of $3+2\epsilon$ from below at the point (2, 3). The form of the graph near (2, 3) is shown in Figure 8.

2.2 The binomial theorem. Before we can use this technique on functions less conveniently defined, we must appeal to the binomial theorem for positive integral indices.

We know that

$$(a+\epsilon)^2 = a^2+2a\epsilon+\epsilon^2 = a^2+2a\epsilon+s_\epsilon$$

$$(a+\epsilon)^3 = a^3+3a^2\epsilon+3a\epsilon^2+\epsilon^3 = a^3+3a^2\epsilon+s_\epsilon$$

$$(a+\epsilon)^4 = a^4+4a^3\epsilon+6a^2\epsilon^2+4a\epsilon^3+\epsilon^4 = a^4+4a^3\epsilon+s_\epsilon$$

and in general $\underline{(a+\epsilon)^N = a^N+Na^{N-1}\epsilon+s_\epsilon.}$

The reader who is not used to handling such results should practice here expanding such expressions as $(2+\epsilon)^3, (3-\epsilon)^2, (-4+\epsilon)^4, (-1+\epsilon)^5$ and so on.

2.3 Form of graph of polynomial. Let us use the results on a numerical example.

Example 1. Examine the form of the graph of $1-12x+2x^3$ near $x = 3$.

As in Section 2.1, we will consider $f(3+\epsilon)$.

We know that

$$f(3+\epsilon) = 1-12(3+\epsilon)+2(3+\epsilon)^3.$$

Using the shortened form of the formulae in Section 2.2, this gives us

$$f(3+\epsilon) = 1-12(3+\epsilon)+2(27+27\epsilon+s_\epsilon)$$
$$= 19+42\epsilon+s_\epsilon.$$

From this we infer that the graph of f touches that of $19+42\epsilon$ (that is, the line through (3, 19) with gradient 42) at (3, 19).

To see in what way it touches we would have to consider in more detail the terms small compared with ϵ. That would involve using the more extended forms of the formulae given in Section 2.2.

Example 2. Examine the form of the graph of x^4-4x^3 in detail near $x = -1$, $x = 3$ and $x = 2$.

$$f(-1+\epsilon) = (-1+\epsilon)^4-4(-1+\epsilon)^3$$
$$= (1-4\epsilon+6\epsilon^2-4\epsilon^3+\epsilon^4)-4(-1+3\epsilon-3\epsilon^2+\epsilon^3)$$
$$= 5-16\epsilon+18\epsilon^2-8\epsilon^3+\epsilon^4.$$
$$f(3+\epsilon) = (3+\epsilon)^4-4(3+\epsilon)^3$$
$$= (81+108\epsilon+54\epsilon^2+12\epsilon^3+\epsilon^4)-4(27+27\epsilon+9\epsilon^2+\epsilon^3)$$
$$= -27+18\epsilon^2+8\epsilon^3+\epsilon^4.$$
$$f(2+\epsilon) = (2+\epsilon)^4-4(2+\epsilon)^3$$
$$= (16+32\epsilon+24\epsilon^2+8\epsilon^3+\epsilon^4)-4(8+12\epsilon+6\epsilon^2+\epsilon^3)$$
$$= -16-16\epsilon+4\epsilon^3+\epsilon^4.$$

The first of these results may be described as normal.

We interpret it by saying the graph of f touches that of $5-16\epsilon$ from above: that is, touches a line with gradient -16 at the point $(-1, 5)$.

The two remaining examples are special cases, though of different kinds. In the first, there is no ϵ term; in the second no ϵ^2 term.

We interpret the first by saying that the graph touches that of $-27+0\epsilon$ from above at $(3, -27)$: that is, it touches a horizontal line from above at this point, so that the graph passes through a minimum at $(3, -27)$.

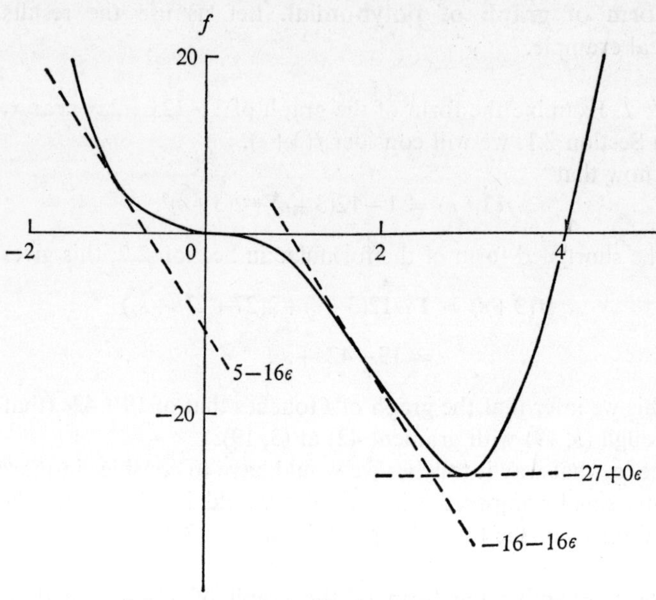

Fig. 9

In the second we do not get a horizontal tangent. The tangent is the graph of $-16-16\epsilon$, which has gradient -16. The graph of f touches this at the point $(2, -16)$, but since it is obtained from that of $-16-16\epsilon$ by adding that of $4\epsilon^3+\epsilon^4$, it touches from above on the right, and from below on the left. Hence the graph of f crosses its tangent at this point: again a special case, though of another kind.

All this information is illustrated in Figure 9.

To determine how the graph lay in relation to its tangents, we considered in each case the terms small compared with ϵ. These were

$$18\epsilon^2-8\epsilon^3+\epsilon^4, \quad 18\epsilon^2+8\epsilon^3+\epsilon^4 \quad \text{and} \quad 4\epsilon^3+\epsilon^4.$$

But the reader may have noticed we took into account only, in effect, the first term in each of these expressions. In the first case we had $18\epsilon^2$, a positive element which gave us the graph touching its tangent from above. In the second case we had the same. In the third we had $4\epsilon^3$, which gave us a curve touching from on top on the right and from below on the left. The reason for this was that, for small ϵ, the terms $18\epsilon^2$ and $4\epsilon^3$ effectively swamped the higher powers of ϵ. We will not here analyse in detail how this happens, beyond saying that just as near $\epsilon = 0$ terms in ϵ^2 are small compared with ϵ, so terms in ϵ^3 are small compared with ϵ^2, and terms in ϵ^4 are small compared with ϵ^3, and so on. Something about this will be found in Exercise A, Questions 14, 15, 16.

Exercise B

1. Sketch the graphs of the functions f near $x = -3$ if it is known that $f(-3+\epsilon)$ is equal to $3+\epsilon-3\epsilon^2$, $2-\epsilon+\epsilon^3$, $\epsilon+\epsilon^2-2\epsilon^5$, $3-\epsilon^4$, $-3\epsilon+\epsilon^5$, $-2-3\epsilon+4\epsilon^2$.

2. Examine the form of the graphs of the following functions of x near the points stated (as in Example 1):

$$2-3x+x^2 \quad \text{near} \quad x = 2,$$

$$3-2x+x^2-x^3 \quad \text{near} \quad x = 1,$$

$$-x^3+3x^2-x \quad \text{near} \quad x = 1,$$

$$x^3-12x \quad \text{near} \quad x = 2,$$

$$x^3+x^4 \quad \text{near} \quad x = -1.$$

3. Find how the graphs in Question 2 lie in relation to the tangents at the relevant points (as in Example 2).

4. Sketch the graph of $3x-x^3$, and examine its behaviour near $x = 0$, $x = 1$, and $x = -1$.

5. Sketch the graph of $2x^3-x^4$ and examine its behaviour near $x = 0$ and $x = 1$.

6*. If $f(x) = x^2-6x+7$, express $f(a+\epsilon)$ in the form $A+B\epsilon+s_\epsilon$. Hence show that the gradient of the tangent to the graph of f is positive, negative or zero according as x is greater than, less than or equal to 3. Sketch it.

7*. Use the method of Question 6 to sketch the graph of $2-x-3x^2$.

8*. What is the form of the graph of x^{1000} near $x = 0$, $x = 1$, $x = 0{\cdot}9$?

3. LINEAR APPROXIMATION

3.1 We have seen that, given a function f, we can quite often express $f(a+\epsilon)$ in the form $A+B\epsilon+s_\epsilon$; and that this enables us to sketch its graph near $(a, f(a))$ in relation to the tangent at this point.

But it does more than this; it gives also an approximate value to $f(a+\epsilon)$ near $\epsilon = 0$. How good an approximation for practical purposes is a question we will not discuss now; though it has considerable theoretical importance.

When we can express $f(a+\epsilon)$ in the form $A+B\epsilon+s_\epsilon$, we will call $A+B\epsilon$ its *linear approximation* near $x = a$.

It is necessary to show that a function can have only one linear approximation (so defined) near a particular point.

Let us suppose that

$$f(a+\epsilon) = A+B\epsilon+s_\epsilon.$$

Putting $\epsilon = 0$, we see that $f(a) = A$. Hence the coefficient A is uniquely defined. Suppose nevertheless there is a second linear approximation, so that

$$f(a+\epsilon) = A+B'\epsilon+s_\epsilon.$$

Subtracting one from the other we get

$$(B-B')\,\epsilon+s_\epsilon = 0 \quad \text{or} \quad (B-B')\,\epsilon = s_\epsilon.$$

But B, B' and hence $B-B'$ are constants. Hence $(B-B')\,\epsilon$ can only be small compared with ϵ if $B-B' = 0$, that is $B = B'$.

It follows that the two linear approximations are identical.

3.2 Existence of linear approximation. This is not a general question that we can discuss here; and in fact it is not true that we can always find a linear approximation. But it is worth remarking that we expect there to be one if the function's graph passes through the point in question in a smooth well-behaved sort of way.

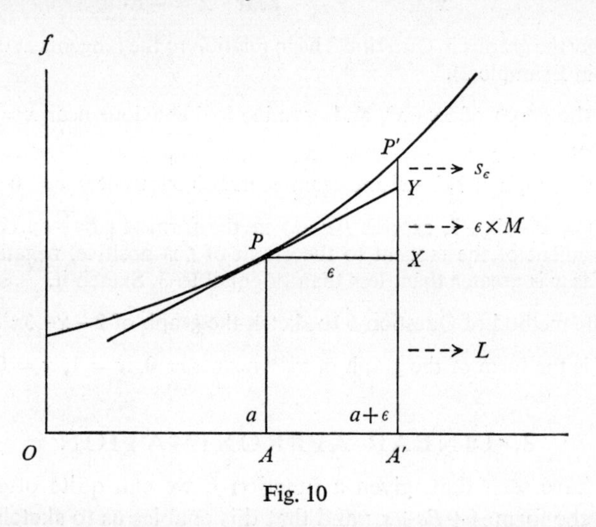

Fig. 10

The informal question to ask is this: can we draw a tangent to the function's graph at $x = a$? If so, then we seem to get a linear approximation near $x = a$.

For the approximation to $f(a+\epsilon)$ (see Figure 10) is measured by YA', and the error by YP'.

Hence
$$f(a+\epsilon) = A'X+XY+YP'.$$

$A'X$ is $f(a)$: L say. XY is $PX\times$(gradient of PY) $= \epsilon\times M$ (say); and from the way in which the graph sidles up to PY, it looks as if YP' is small compared with ϵ.

Hence
$$f(a+\epsilon) = L+M\epsilon+s_\epsilon,$$

where (as we have seen) L is $f(a)$, and M is the gradient of the tangent at P.

$L+M\epsilon$ is, then, the linear approximation to $f(x)$ near $x = a$. This 'proof' is quite bogus. It depends on the assumption that we can draw a tangent. But what is a tangent? If we look back to Section 1.3, we see that our definition of tangency implied the existence of an error (here PY') small compared with PX, or ϵ.

It may however help to make clear the condition under which we expect there to be a linear approximation.

Exercise C

1. Find the linear approximations to the following functions of x near the stated values:

$$4-2x^2 \quad \text{near} \quad x = 3,$$

$$-2+7x-x^3 \quad \text{near} \quad x = -1,$$

$$3x^2-x^5 \quad \text{near} \quad x = 2,$$

$$1+\tfrac{1}{4}x+\tfrac{1}{2}x^4 \quad \text{near} \quad x = -\tfrac{1}{2}.$$

2. Show that the linear approximation to x^3 near $x = 2$ is $8+12\epsilon$ with error $6\epsilon^2+\epsilon^3$.

For which of the following values of ϵ is the error less than 0·25? $\epsilon = 0{\cdot}01, 0{\cdot}1, 0{\cdot}2, 0{\cdot}3$. For roughly what values of ϵ is the error less than 0·25? Find the cube of 2·15 to this accuracy. For roughly what values of ϵ is the error less than 10^{-4}? Find the cube of 1·995 to this accuracy.

Roughly how accurate in the linear approximation in finding the cube of 1·973?

For roughly what values of ϵ is the error smaller than $0{\cdot}01\epsilon$?

4. THE DERIVED FUNCTION

4.1 Derivatives. If $A+B\epsilon$ is a linear approximation to $f(a+\epsilon)$,

$$f(a+\epsilon) = A+B\epsilon+s_\epsilon.$$

Putting $\epsilon = 0$, we get $A = f(a)$. But what is B?

We have noted of course that B is always the gradient to the tangent to the function's graph at $(a, f(a))$; but this puts the answer in merely geometrical terms, and B has a wider significance than this.

We shall see that significance more fully in Chapter 5; but even here it is useful to give a name to the coefficient of ϵ in the linear approximation. We shall call it the *derivative* of $f(x)$ where $x = a$.

The derivative then, where $x = a$, measures the gradient of the tangent at the point $(a, f(a))$. It distinguishes, for instance, whether the graph is rising (positive derivative) or falling (negative). It measures also how fast it is rising, or how steeply (the bigger the derivative the steeper); tells us whether the graph is passing through a highest or lowest point (where the derivative will be zero) and so on. But we can go further than this.

We note in the first place that, for functions of the kind we have met so far, derivatives can be calculated for all elements of the domain. Take for example $f(x) = 3x^2 - x^3$. Figure 11 shows the sketch of its graph, obtained by the methods of Chapter 2. We would expect there to be a tangent at every point, and hence a linear approximation. Therefore we expect a derivative of $f(x)$ for each value of x.

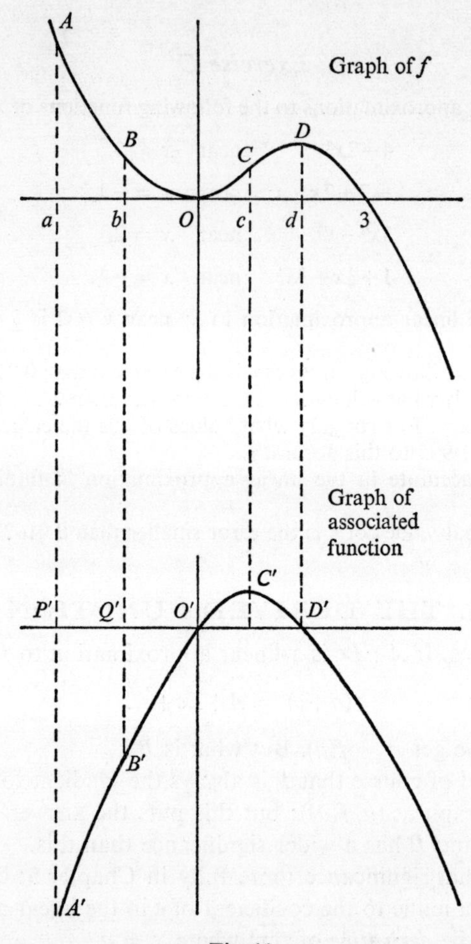

Fig. 11

It seems then that, associated with f, there is another function which maps x onto the derivative of f(x) for all values of x.

We can even sketch this function's graph from that of f. (See the lower half of Figure 11.)

For consider the graph of f. For $x < 0$ the tangents slope downwards; hence the derivatives must be negative—numerically large at A (where the tangent is steepest) getting smaller and approaching zero as we move through B to O.

Hence as x moves from $-\infty$ to 0, the derivative of $f(x)$ moves from $-\infty$ to 0. This is shown in the section $A'O'$ of the lower graph: the length $A'P'$ corresponding to the derivative where $x = a$, $B'Q'$ to the derivative where $x = b$, and so on.

Consider now the section OC. Here the derivative increases from 0 to a maximum value where $x = c$ ($O'C'$ in the lower graph). Then as x increases from c to d, the derivative again falls to zero ($C'D'$). And as x increases from d through 3 to ∞, the derivative again becomes negative and falls from 0 to $-\infty$.

Hence the graph of the associated function is as shown, though we cannot yet determine the values of c and d (where the derivative is respectively maximum and zero).

Let us check this picture by calculating at random a few derivatives.

$$f(-1+\epsilon) = 3(1-2\epsilon+s_\epsilon)-(-1+3\epsilon+s_\epsilon) = 4-9\epsilon+s_\epsilon;$$

so that the derivative at $(-1, 4)$ is -9: a reasonable looking result.

$$f(1+\epsilon) = 3(1+2\epsilon+\epsilon^2)-(1+3\epsilon+3\epsilon^2+\epsilon^3) = 2+3\epsilon-\epsilon^3.$$

Here we have worked out also the error term in the linear approximation, and it reveals that the graph crosses its tangent at this point, giving a maximum derivative of 3 where $x = 1$. It follows that $c = 1$.

$$f(4+\epsilon) = 3(16+8\epsilon+s_\epsilon)-(64+48\epsilon+s_\epsilon) = -16-24\epsilon+s_\epsilon,$$

so that the derivative at $(4, -16)$ is -24.

The drill, we note, is simple. The first element in the linear approximation $(4, 2, -16)$ is the value of $f(x)$. The second element, or coefficient of ϵ, is the value of the derivative.

All this suggests that we might find a general value of the derivative by examining $f(a+\epsilon)$.

$$f(a+\epsilon) = 3(a+\epsilon)^2-(a+\epsilon)^3 = 3(a^2+2a\epsilon+s_\epsilon)-(a^3+3a^2\epsilon+s_\epsilon)$$
$$= (3a^2-a^3)+(6a-3a^2)\,\epsilon+s_\epsilon.$$

Hence the derivative of $f(x) = 3x^2-x^3$ at $x = a$ is $6a-3a^2$. This checks and completes the picture we have so far obtained.

When $a = -1$, the derivative is $-6-3 = -9$.

When $a = 4$, the derivative is $24-48 = -24$.

Lastly we can calculate that the derivative is zero where

$$6a-3a^2 = 3a(2-a) = 0,$$

i.e. $$a = 0 \quad \text{or} \quad 2.$$

This verifies that we have a zero derivative where $x = 0$ and further shows a zero derivative (or that the graph of f passes through a highest point) where $x = 2$. This last is a new piece of information and completes our picture of the graph.

The second function, mapping x onto the derivative of $f(x)$, is called the *derived function* of f. It is written f'. In full:

If a function of x, $f(x)$, has a derivative at every value of x (for some interval), we call the function which maps x onto this derivative the derived function of f over that interval. We write this derived function f' so that the derivative for any value of x may be written $f'(x)$.

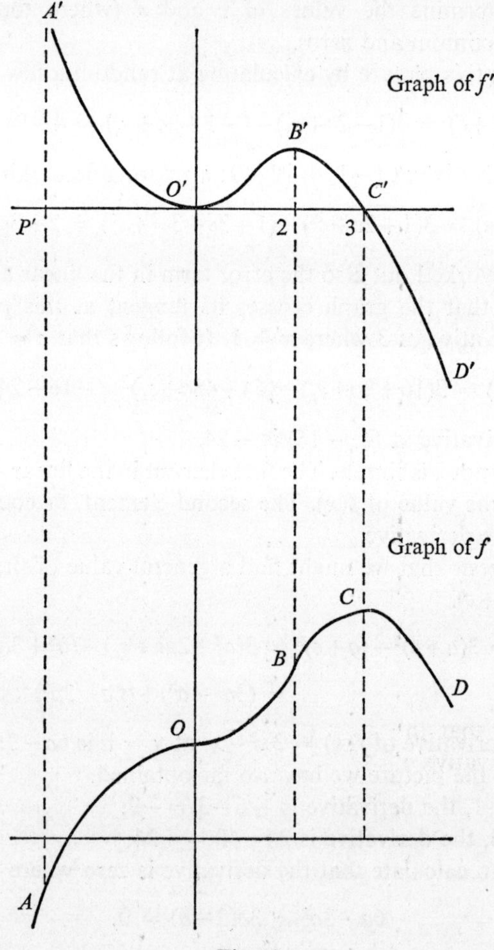

Fig. 12

This is $f(x) = 3x^2 - x^3$, since the derivative of $f(x)$ at $x = a$ is $6a - 3a^2$,

$$f'(a) = 6a - 3a^2 \quad \text{and} \quad f'(x) = 6x - 3x^2.$$

The derived function is therefore, in this case, the quadratic whose graph is shown in the lower half of Figure 11.

4.2 Graph of function from that of derived function. The highly important relationship between the graphs of f and f' is well seen by putting the last problem into reverse. If we know f' and the appearance of its graph, can we sketch roughly the graph of f?

Suppose this time that $f'(x) = 3x^2 - x^3$. From its graph (the upper half of Figure 12) we observe that as x increases from $-\infty$ to 0, the derivative of $f(x)$ (measured by lengths such as $A'P'$) falls from ∞ to 0. This could correspond to a stretch of graph looking like AO in Figure 12, whose gradient falls from large values (as at A) to zero at O.

As x now increases from 0 to 2, $f'(x)$ rises again to a maximum value (OB in the graph of f). As x increases from 2 to 3, $f'(x)$ again falls to zero (BC); and thereafter, as x increases to ∞, $f'(x)$ falls to $-\infty$ (CD and onwards).

4.3 Linear approximation. The notation of derived function allows us to give the linear approximation its final form. The gradient of the tangent in Figure 13 is $f'(a)$.

Hence the length XY is $f'(a).\epsilon$.

The length $A'X$ is $f(a)$.

The length YP' is s_ϵ.

Thus
$$\begin{aligned} f(a+\epsilon) &= A'X + XY + YP' \\ &= f(a) + f'(a).\epsilon + s_\epsilon. \end{aligned}$$

We infer this from the graph, assuming that it has a tangent at $(a, f(a))$.

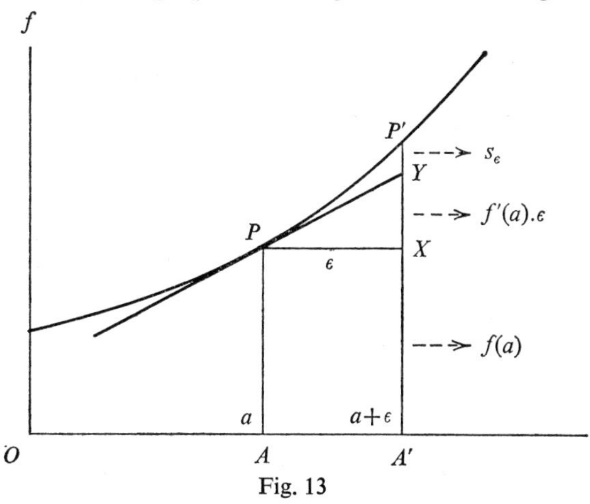

Fig. 13

Otherwise, if we find from algebraic analysis that, for example,

$$f(5+\epsilon) = -2+\tfrac{1}{4}\epsilon+s_\epsilon,$$

we deduce that $$f(5) = -2,$$

that the gradient at $(5, -2)$ is $\frac{1}{4}$, or that $f'(5) = \frac{1}{4}$.

Hence again $$f(5+\epsilon) = f(5)+f'(5).\epsilon+s_\epsilon.$$

This is the standard form of the linear approximation which in future we shall always quote.

Exercise D

1. Draw accurately the graph of $f(x)$ from the table of values:

x =	-2	$-1\cdot5$	-1	$-0\cdot5$	0	$0\cdot5$	1	$1\cdot5$
$f(x)$ =	28	$11\cdot0$	3	$0\cdot3$	0	$-0\cdot3$	-2	$-5\cdot9$
x =	2	$2\cdot5$	3	$3\cdot5$	4	$4\cdot5$	5	$5\cdot5$
$f(x)$ =	-12	$-19\cdot5$	$-2\cdot7$	$-32\cdot2$	-32	$-25\cdot9$	0	$41\cdot6$

Estimate the gradients of the graph's tangents at $x = -2, -1, 0, 1, 2, 3, 4, 5$, and draw the graph of the derived function f' from this data.

2. In the figure are shown the graphs of a number of functions. Copy them, and below (as in Section 4.1) sketch the graphs of their derived functions, showing clearly in each case the connections between the graphs of function and derived function:

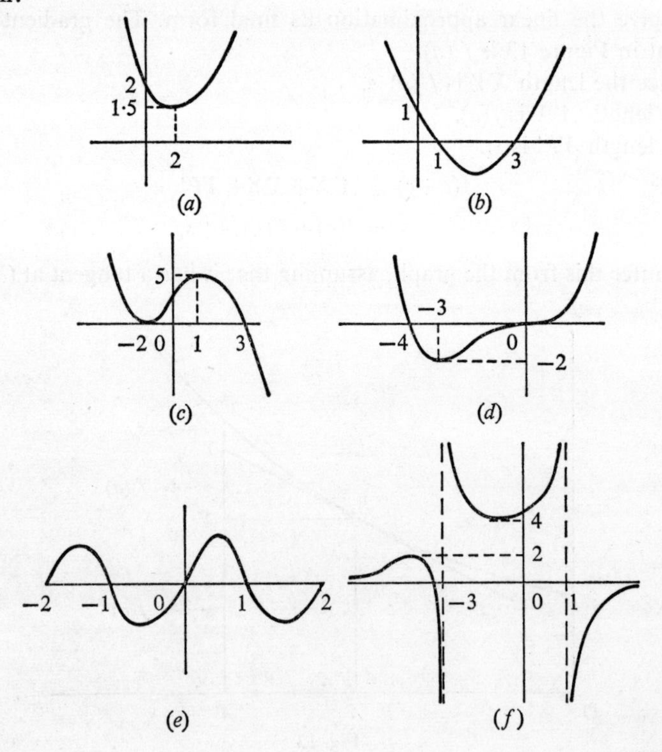

3. In the figure are shown the graphs of a number of derived functions. Copy them and below (as in Section 4.2) sketch the graphs of functions from which they might be derived, showing clearly in each case the connections between the graphs of derived function and function:

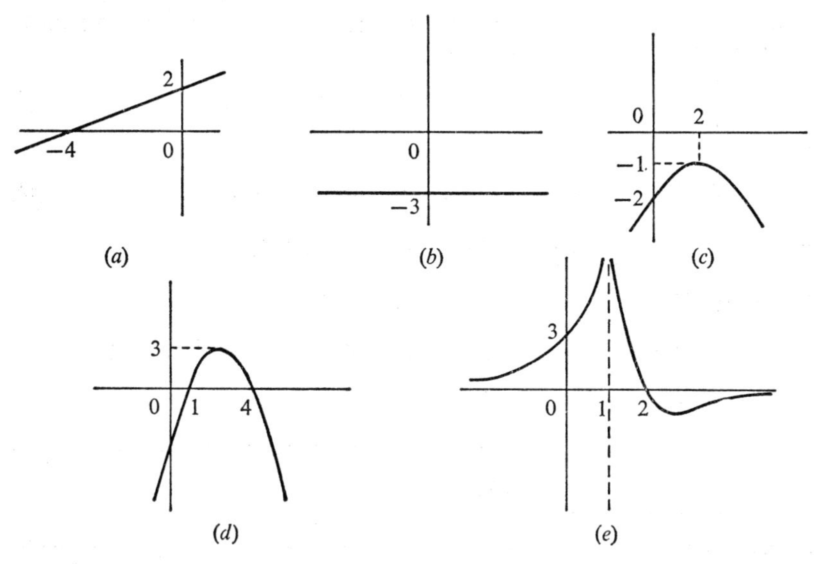

4. What are the derived functions of those functions which map x always onto $5x$, $-3x$, 7?

What can you say about the functions whose derived function map x always onto 5, -2, 0?

5. Sketch the graph of $f(x) = x^2-3x$. Hence sketch the graph of the derived function. What can you say about the derived function? Check by calculating $f'(a)$ as in Section 4.1.

6. From the graph of x^2 sketch the graph of its derived function. What can you say about the derived function? Check by calculating it for $x = a$.

7. Repeat Question 6, (*a*) for x^3, (*b*) for x^4.

8. Repeat Question 5, (*a*) for $f(x) = x^3-3x^2+5x$ (in sketching the graph of f add those of x^3-3x^2 and $5x$); and (*b*) for $f(x) = 2x^3+x^4$.

9. Draw the graphs of x^3+2x and x^3-2x. Sketch the graphs of their derived functions, and check by calculating them in both cases.

10. $f'(x) = 3x^2-6x+5$. Sketch its graph, noting that it does not dip below the x-axis, and sketch the general shape of the graph of f if you know that $f(0) = -3$. Why does the last piece of information help?

11. Repeat Question 10 for $f'(x) = 3x^2-6x+2$.

12. Repeat Question 10 for (*a*) $f'(x) = x^{-2}$, $f(1) = 1$; (*b*) $f'(x) = x^{-1}$, $f(1) = 0$.

5. DERIVATIVES

5.1 Calculating derivatives. For polynomial functions, we can calculate derivatives after the manner of Section 4.1. But this is tedious; and the careful reader may have noticed that we can do it more easily once we know the derivatives of simple powers of x.

Now if
$$f(x) = x^2, \quad f(a+\epsilon) = a^2+2a\epsilon+s_\epsilon. \qquad \text{(i)}$$
Hence $\quad f'(a) = 2a \quad \text{or} \quad f'(x) = 2x.$

Likewise if
$$f(x) = x^3, \quad f(a+\epsilon) = a^3+3a^2\epsilon+s_\epsilon. \qquad \text{(ii)}$$
Hence $\quad f'(a) = 3a^2 \quad \text{or} \quad f'(x) = 3x^2.$

In general, if $\quad f(x) = x^N \quad \text{for} \quad N > 1,$
$$f(a+\epsilon) = (a+\epsilon)^N = a^N+Na^{N-1}\epsilon+s_\epsilon \quad \text{(see Section 2.2);}$$
hence $\quad \underline{f'(x) = Nx^{N-1}.}$

$f(x) = x = x^1$ also obeys this rule, for its derivative is always 1 and, by the formula, $f'(x) = 1.x^0 = 1$.

Likewise if $f(x) = 1 = x^0, f'(x) = 0.x^{-1} = 0$: also a true result. Thus we have a formula for all non-negative integers N. It can also be put in the form:
$$P_2' = 2P_1, \quad P_3' = 3P_2;$$
and in general $\quad P_N' = NP_{N-1}.$

Now consider how we might deal with more complicated functions, like
$$f(x) = 5x^2 \quad \text{or} \quad g(x) = 5x^2-2x^3.$$
$$f(a+\epsilon) = 5(a+\epsilon)^2 = 5(a^2+2a\epsilon+s_\epsilon)$$
$$= 5a^2+5\times 2a.\epsilon+s_\epsilon;$$
and
$$g(a+\epsilon) = 5(a+\epsilon)^2-2(a+\epsilon)^3 = 5(a^2+2a\epsilon+s_\epsilon)-2(a^3+3a^2\epsilon+s_\epsilon)$$
$$= (5a^2-2a^3)+(5\times 2a-2\times 3a^2)\,\epsilon+s_\epsilon.$$
We see in fact, comparing these with the equations (i) and (ii) above, that we get the derivative of $5x^2$ by writing $5\times(\text{derivative of } x^2) = 5\times 2x = 10x$; and the derivative of $5x^2-2x^3$ by writing
$$5\times(\text{derivative of } x^2)-2\times(\text{derivative of } x^3)$$
$$= 5\times 2x \qquad -2\times 3x^2$$
$$= 10x-6x^2.$$

These facts may be formulated into the following rules:

I if $g(x) = kf(x)$ then $g'(x) = kf'(x)$ if k is a constant;

II if $h(x) = g(x)+f(x)$ then $h'(x) = g'(x)+f'(x)$;

and their consequence:

III if $h(x) = mf(x)+ng(x)$,

where m and n are constants, then

$$h'(x) = mf'(x)+ng'(x).$$

The drill therefore for differentiating polynomials is an extremely simple one. We have the fundamental table:

$f(x) = 1\ (=x^0)$	$x\ (=x^1)$	x^2	x^3	x^4	x^5	...
$f'(x) = 0$	1	$2x$	$3x^2$	$4x^3$	$5x^4$	...

Thus when we *differentiate* functions (that is, obtain derived functions) we get results like the following:

$7x^3$ becomes $7\times 3x^2 = 21x^2$

x^2+x^5 becomes $2x+5x^4$

$2x^6-3x^2-x+17$ becomes $2\times 6x^5-3\times 2x-1\times 1+17\times 0$

$$=12x^5-6x-1.$$

5.2 Graph sketching. With this improved ability to differentiate, let us return to the problem of graph sketching. For a knowledge of the derived function enables us to settle various points that were obscure before: where, for instance, $f(x)$ passes through its greatest and least values, or where the gradient is greatest. Sometimes also it is easiest to sketch first the graph of $f'(x)$ and guess the form of $f(x)$ from it.

Example 1. Sketch the graph of x^2-x+1. (See Figure 14.)

Here we can sketch the graph by adding those of x^2 and $1-x$. The methods of Chapter 2 do not, however, settle where the minimum value occurs, nor whether the graph dips below the x-axis: an important point, which will determine whether $x^2-x+1 = 0$ has any solutions.

Let us therefore differentiate, getting

$$f'(x) = 2x-1.$$

We see that $f'(x) = 0$ if $x = \frac{1}{2}$.

Hence the minimum point occurs where $x = \frac{1}{2}$. But $f(\frac{1}{2}) = \frac{1}{4}-\frac{1}{2}+1 = \frac{3}{4}$. The minimum value is, therefore, positive, and $f(x)$ can never be zero or negative.

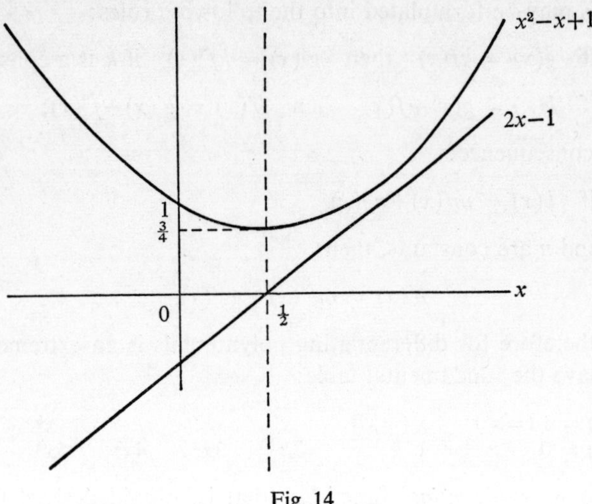

Fig. 14

Another point is suggested by sketching the graph of $f'(x)$ below that of $f(x)$: that the graph of f is symmetrical about the vertical axis through $(\frac{1}{2}, 0)$; since at equal intervals on either side of this line the gradients (obtained from the derivative $2x-1$) are numerically equal and opposite in sign. This symmetry is a characteristic of the graphs of all quadratic functions.

Example 2. Sketch the graph of

$$f(x) = x^3-3x^2+5x-3.$$

Here again we could add the graphs of x^3-3x^2 and $5x-3$. It is, however, rather easier to sketch first the graph of f', and infer the nature of that of f from it.

$$f'(x) = 3x^2-6x+5.$$

Its graph can be sketched in the same way as that in Example 1 and is shown in the upper half of Figure 15.

The graph of f is shown below it. The gradient is least when $x=1$; and $f(1)$ is equal to 0.

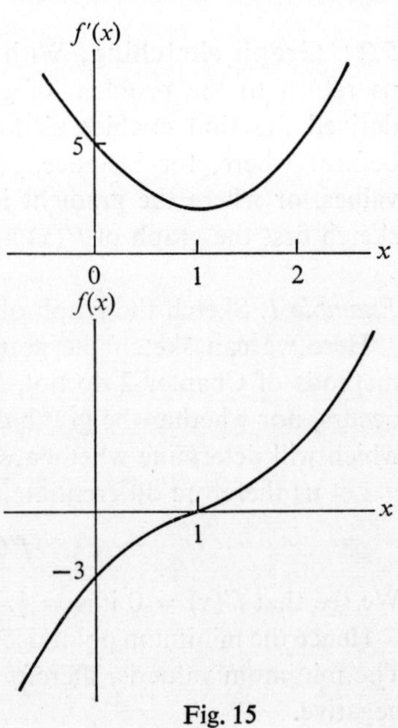

Fig. 15

5.3 A note on terminology. So far in this chapter we have met the words *derivative*, *derived function* and *differentiate*. These can be confusing if carelessly used. In this book we shall use them consistently in the following senses:

> the activity of obtaining f' from f *or* $f'(x)$ from $f(x)$ we describe as *differentiating*; thus we *differentiate* P_3 to obtain $3P_2$; and we *differentiate* $3x^2-x$ to obtain $6x-1$.
>
> The result of differentiating a *function* we call its *derived function.* Thus again f' is the derived function of f, and $3P_2$ of P_3. The result of differentiating a *function of* x we call its *derivative.* Thus $3x^2$ is the derivative of x^3, and $f'(x)$ the derivative of $f(x)$.

It will be observed from this that a derived function is a function (as its name implies); and a derivative is a variable or number. We would not speak of $3x^2$ as a derived function, since it is not a function but a variable. And we would not speak of f' as a derivative, since it is not a variable but a function.

Exercise E

1. Write down the derivatives of: $7x$, $3x^2$, $5x-2$, $2x-x^3$, $2+x^7$, $2-x+3x^3-7x^6$.

2. Find the gradients to the graphs of the following functions of x at the points stated:

$$\tfrac{1}{4}x^4-3x^3 \quad \text{at} \quad x=0 \quad \text{and at} \quad x=2$$
$$x(x+7) \quad \text{at} \quad x=0 \quad \text{and at} \quad x=-1$$
$$x(x-1)(2x+1) \quad \text{at} \quad x=-2.$$

(In the last two cases, multiply out the expressions before differentiating.)

3. If $f(x) = 2x^4+7x^2$, what are $f'(2)$ and $f'(-1)$? For what values of x is $f'(x) = 0$?

4. For what values of x does $3x^2-2x$ decrease as x increases?

5. Sketch the graphs of the following functions of x by the methods of Chapter 2, and then use their derivatives to get as much more information as you can:

$$x^2-3x, \quad x^2+4x+2, \quad x^3-x, \quad 2x^2+3x^3, \quad 4x^3-x^4.$$

6. Sketch the graphs of the following functions of x by first sketching the graphs of their derivatives:

$$x^2+4x+5, \quad x^2+4x+1, \quad 6x-x^3, \quad 6x+x^3.$$

7. Sketch the graph of x^2-2x+2. Hence sketch the graph of

$$x^3-3x^2+6x-10.$$

8. Sketch the graph of x^2-2x-1. Hence sketch the graph of

$$x^3-3x^2-3x-4.$$

9. Sketch the graph of $12x^2-4x^3$. Hence sketch that of $-7x+4x^3-x^4$.

10. Sketch the graph of $3x^4+4x^3+\tfrac{3}{2}x^2+1$.

6. THE RECIPROCAL FUNCTION

We shall finish this chapter by considering briefly one function which requires a little more subtlety in finding a linear approximation.

6.1 Derivative of x^{-1}. Suppose $f(x) = x^{-1}$. We wish to find a linear approximation: that is obtain an equation of the form:

$$f(a+\epsilon) = f(a) + B\epsilon + s_\epsilon$$

or
$$\frac{1}{a+\epsilon} = \frac{1}{a} + B\epsilon + s_\epsilon$$

so that we can identify B with the derivative of $f(x)$ at $x = a$.

We do this most easily by long division.

$$\begin{array}{r|l}
 & \dfrac{1}{a} - \dfrac{\epsilon}{a^2} \\ \hline
a+\epsilon & 1 \\
 & 1 + \dfrac{\epsilon}{a} \\ \cline{2-2}
 & -\dfrac{\epsilon}{a} \\
 & -\dfrac{\epsilon}{a} - \dfrac{\epsilon^2}{a^2} \\ \cline{2-2}
 & \dfrac{\epsilon^2}{a^2}
\end{array}$$

We see then that
$$\frac{1}{a+\epsilon} = \frac{1}{a} - \frac{1}{a^2}\epsilon + \frac{\epsilon^2}{a^2(a+\epsilon)}.$$

Now the last term is small compared with ϵ. For anything of the form $\epsilon^2 g(\epsilon)$ will be small compared with ϵ where $g(\epsilon)$ is well behaved and finite near $\epsilon = 0$. (See Exercise A, Question 10. An example of a *non*-well behaved function here would be one that tends to ∞ as ϵ tends to 0. For instance ϵ^{-1}: $\epsilon^2\epsilon^{-1}$, obviously, is not small compared with ϵ.)

Hence we can write
$$\frac{1}{a+\epsilon} = \frac{1}{a} - \frac{1}{a^2}\epsilon + s_\epsilon$$

or
$$f(a+\epsilon) = f(a) - \frac{1}{a^2}\epsilon + s_\epsilon,$$

so that
$$f'(a) = -\frac{1}{a^2}.$$

This conforms with the theorem $f(x) = x^N \Rightarrow f'(x) = Nx^{N-1}$ which was proved for positive integral values of N in Section 5.1; for

$$f(x) = x^{-1} \Rightarrow f'(x) = -1x^{-2}.$$

Example: Sketch the graph of $f(x) = \dfrac{x^3-16}{x}$.

Both to differentiate and to sketch the graph more easily, we do best to write:

$$f(x) = x^2 - 16x^{-1}.$$

We can sketch the graph by adding the graphs of x^2 and $-16x^{-1}$ (shown by dotted lines in Figure 16). Clearly we get a minimum for some negative value of x.

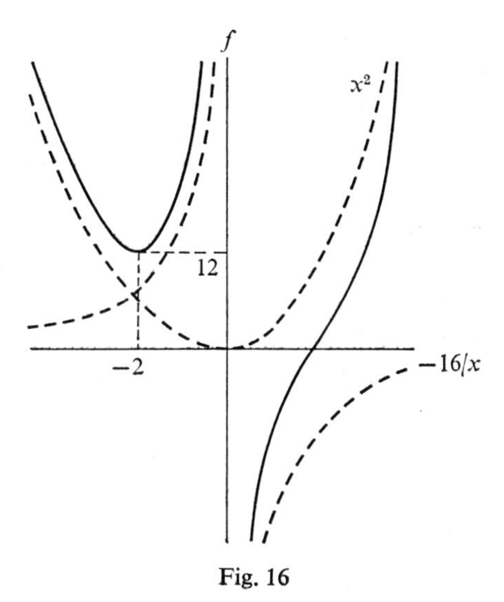

Fig. 16

$$f'(x) = 2x + 16x^{-2} = 0$$
if
$$x^3 = -8, \quad \text{or if} \quad x = -2.$$

This gives us the minimum.

$$f(-2) = 4+8 = 12.$$

Hence Figure 16.

Exercise F

1. What are the derivatives of:

$$3x^2+7x^3-\frac{2}{x}, \quad x+\frac{1}{x}, \quad \frac{7x^3-5}{2x}?$$

2. Find the derivatives of the following functions of x at the stated values of x:

$$\tfrac{1}{2}x^2-3x^3+\frac{2}{x} \text{ at } x=2; \quad \frac{x^3}{6}-\frac{4}{x} \text{ at } x=-\tfrac{1}{2}.$$

3. Sketch the graph of $f(x) = x+\dfrac{4}{x}$, and that of $f'(x)$, showing clearly the relation between them.

4. Repeat Question 3 for $\quad f(x) = 4x^2+\dfrac{1}{x}$.

5. Repeat Question 3 for $\quad f(x) = \dfrac{2x+3}{x}$.

4

APPLICATIONS OF GRAPH SKETCHING

In this chapter we turn to some of the simpler uses of graph sketching and of the derived function. They are at this stage two: discussing the solution of equations, and physical problems in which we can express one variable as a function of another. The methods we shall describe here are of wide application; but at the moment we can illustrate them only with simple examples.

1. SOLUTION OF EQUATIONS

1.1 Graphical method. From a sketch of a function f, it is often possible, without much computation, to decide how many values of x make $f(x)$ zero and to determine these values approximately. These values of x are called the 'zeros of the function' or the 'roots' of the equation $f(x) = 0$, and the process of finding them is called 'solving the equation'.

The approximate solution of the cubic equation in the next example will be found to depend on a few easily computed values of $f(x)$.

Example. Solve $x^3-3x^2-45x+200 = 0$ to the nearest integer.

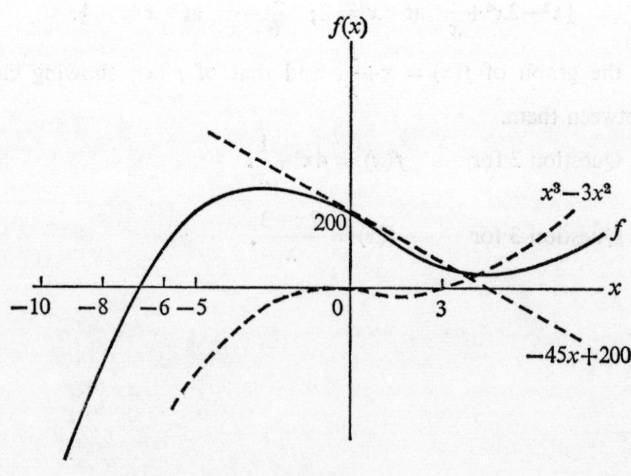

Fig. 1

Let $f(x) = x^3-3x^2-45x+200$. The form of the graph of f can be found by adding the graphs of x^3-3x^2 and $-45x+200$, as shown in Figure 1. Having noted the scales chosen, we can see that the graph of f cuts the x-axis at only one point—that is, f has only one zero.

Now $f(-10)$ is roughly -1000, so that the root lies between 0 and -10. But

$$f(-5) = -125-75+225+200 = 225,$$

which shows, when the points corresponding to $x = -5$ and $x = -10$ are plotted, that the root almost certainly lies between -6 and -8.

The values of $f(-6)$ and $f(-8)$ are easily calculated: $f(-6) = 146$ and $f(-8) = -144$. From these values it is clear that the root is close to -7.

When a rough approximation to a root has been found, there is a simple routine, attributed to Newton, for finding a closer approximation.

1.2 Newton's method. Suppose we know that there is a root of the equation $f(x) = 0$ near $x = p$, and we are looking for a closer approximation to it. In Figure 2, the root is represented by the point R where the graph of f cuts the x-axis. T is the point of the graph corresponding to $x = p$; and the tangent to the graph at T cuts the x-axis at Q.

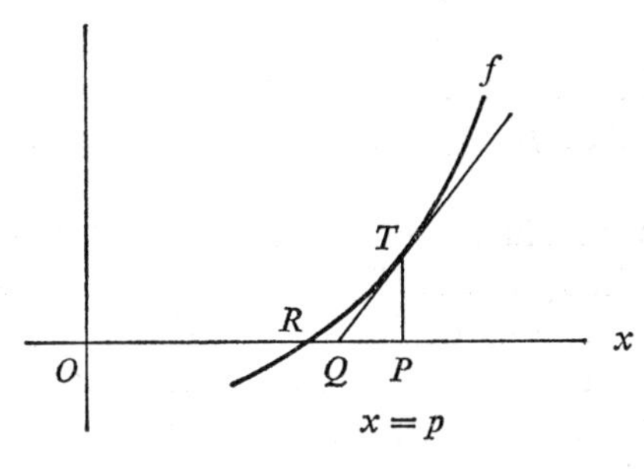

Fig. 2

Now provided P is close enough to the root, we may expect Q to be even closer. Figure 2 shows a case where this is obviously true. Figure 3 on the other hand shows one where Q is not closer. But if in Figure 3 we had started from a better approximation, P' say, it looks again as if Q' would have given a better one still.

To find the value of x at Q, we determine where the linear approximation to $f(x)$ at T is zero. This linear approximation is given by

$$f(p+\epsilon) = f(p)+\epsilon f'(p),$$

and this is zero where $\epsilon = -f(p)/f'(p)$.

Hence, at Q, $x = p+\epsilon = p-f(p)/f'(p)$.

This may be argued in a slightly different way. $TP = f(p)$, and therefore $QP = f(p)/f'(p)$, since the gradient of QT is $f'(p)$. Therefore at Q, $x = p - f(p)/f'(p)$ as before.

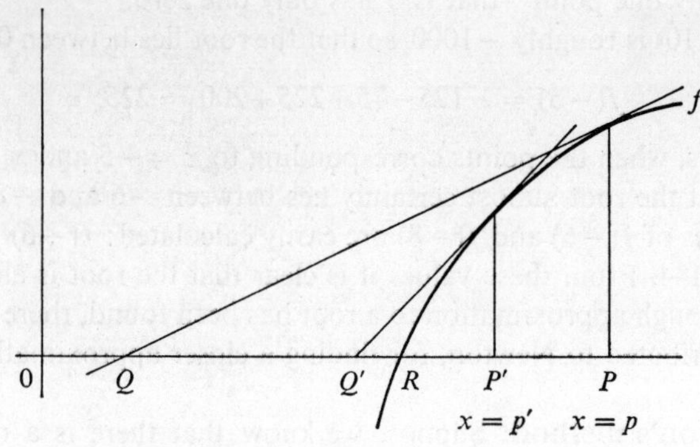

Fig. 3

Hence we can say that if $x = p$ is an approximate root of $f(x) = 0$, a closer approximation will often be

$$x = p - \frac{f(p)}{f'(p)}.$$

In the example in Section 1.1, the root is near $x = -7$, so that $p = -7$. It will be found that $f(-7) = 25$, and $f'(-7) = 144$; and therefore $f(p)/f'(p) = 0{\cdot}17$.

Hence a closer approximation may be $-7-0{\cdot}17 = -7{\cdot}17$.

From the form of the graph in Figure 1, it can be seen that the root is not less than this. By computing we find that $f(-7{\cdot}16) = +1{\cdot}4$. Thus the root can be shown to lie between $-7{\cdot}16$ and $-7{\cdot}17$. Near the root, $f'(x)$ is very approximately 144, so that the root is nearer $-7{\cdot}16$. We can therefore state that the root is $-7{\cdot}16$ to two decimal places.

In this example $f'(x)$ is large in the neighbourhood of the root and this has helped to give a good approximation quickly. In other cases it may be necessary to use Newton's approximation more than once.

Example. Use Newton's method to find the cube root of 10.

Here we wish to solve the equation

$$f(x) = x^3 - 10 = 0.$$

If our first approximation is $x = p_1$, and the second $x = p_2$, then

$$p_2 = p_1 - f(p_1)/f'(p_1)$$
$$= p_1 - (p_1^3 - 10)/3p_1^2,$$

which we can simplify to

$$p_2 = (2p_1^3 + 10)/3p_1^2.$$

Suppose we use this formula to get a chain of successive approximations

$$p_1, p_2, p_3, \ldots, p_r, \ldots.$$

Then in general we can use the same algebra to prove that

$$p_{r+1} = (2p_r^3+10)/3p_r^2.$$

The integer whose cube is nearest to 10 is 2. Let us take this as p_1. The calculation arising from the formula is best set out in a table as below:

r	p_r	p_r^2	p_r^3	$2p_r^3+10$	$3p_r^2$	$(2p_r^3+10)/3p_r^2 = p_{r+1}$
1	2	4	8	26	12	2·17
2	2·17	4·709	10·22	30·44	14·13	2·154
3	2·154	4·6397	9·994	29·998	13·919	2·1545

By computation we find that

$$(2{\cdot}1545)^3 = 10{\cdot}0009089,$$

$$(2{\cdot}1544)^3 = 9{\cdot}999518.$$

Hence the cube root of 10 is 2·1544 (to 4 decimal places).

Exercise A

1. Sketch the graph of $f(x) = x^3-3x+3$, finding the turning values. Hence show that $f(x) = 0$ has only one root, and estimate it correct to 1 decimal place.

2. Solve to one decimal place the following equations:

$$x^3-12x-34 = 0; \quad x^3-7x^2-12 = 0.$$

3. If p_r is an approximation to the square root of 10, show that, by Newton's method, the next approximation is

$$p_{r+1} = (p_2^2+10)/2p_r.$$

Hence find the square root of 10 correct to two decimal places.

4. Solve to two decimal places the following equations:

$$x^5+x-4 = 0; \quad x^4-4x^2+2x-5 = 0.$$

5. State the number of positive and negative roots of:

(i) $2x^3+6x^2-2x-3 = 0$, by sketching the graphs of $2x^3+6x^2$ and $2x+3$,

(ii) $x^4-9x^2+x+1 = 0$,

(iii) $x^4-9x^2+x-1 = 0$,

(iv) $x^5+10x^4-27x^2+12 = 0$.

6. Express $\dfrac{x+1}{x-1} = 2x^2-3$ as a polynomial equation and show that there exists a root between 1 and 2. Find an interval for x, of length 0·1, within which this root lies.

1.3 Solubility of equations. We show in this section two methods for discussing the solubility of certain types of equation: whether for instance and under what conditions we expect then to have 0 or 1 or 2 or more roots.

Example 1. Discuss the solubility of $x^3-3x+k = 0$, for different ranges of value of the constant k.

The equation can be transformed into

$$k = 3x-x^3 = x(3-x^2).$$

Suppose $f(x) = x(3-x^2)$. The graph of f is shown in the figure.

$$f'(x) = 3-3x^2$$

and therefore, $\quad f'(x) = 0, \quad \text{where} \quad x = \pm 1.$

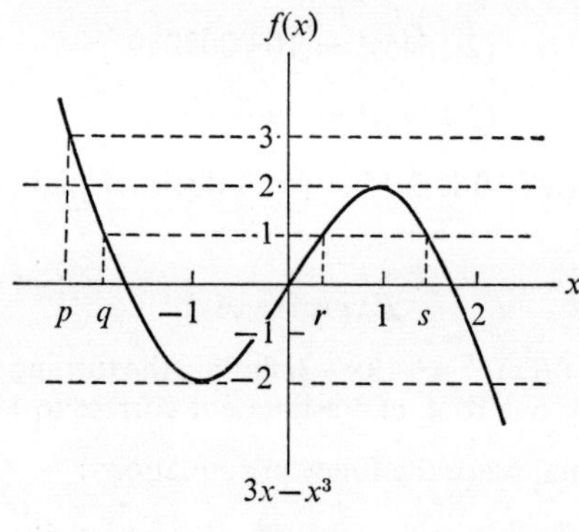

Fig. 4

It is easily checked that where $x = 1$, $f(x)$ has the maximum value 2, and where $x = -1$, $f(x)$ has the minimum value -2. Now, in particular, it can be seen from the nature of the graph that if $x = p$, $3x-x^3 = 3$, and there is no other value of x for which $3x-x^3 = 3$.

Therefore, $3x-x^3 = k$ has only one solution for $k = 3$.

Again, if x has any of the values q, r or s, $3x-x^3 = 1$, and these are the only values of x for which $3x-x^3 = 1$.

Therefore $3x-x^3 = k$ has three solutions if $k = 1$.

In general, arguing in the same way, it can be seen that

if $k > 2$ or $k < -2$, $3x-x^3 = k$ has only one solution;
if $k = 2$ or -2, $3x-x^3 = k$ has two solutions;
and if $-2 < k < 2$, $3x-x^3 = k$ has three solutions.

Example 2. Discuss the solubility of the equation $x^3-3kx+2 = 0$ for different ranges of values of the constant k.

Here we could recast the equation in such a way as to isolate the constant k.

$$x^3-3kx+2 = 0$$

$$\Rightarrow 3kx = x^3+2$$

$$\Rightarrow k = (x^3+2)/3x = \frac{x^2}{3}+\frac{2}{3x}.$$

We could sketch the graph of this function of x, calculating the turning values and arrive at an answer in the same way as in Example 1.

Alternatively we can write

$$kx = (x^3+2)/3$$

and sketch the graphs of kx and $(x^3+2)/3$ as in Figure 5. The graph of kx is a line through the origin of gradient k. Where it cuts the graph of $(x^3+2)/3$, we have roots of the original equation.

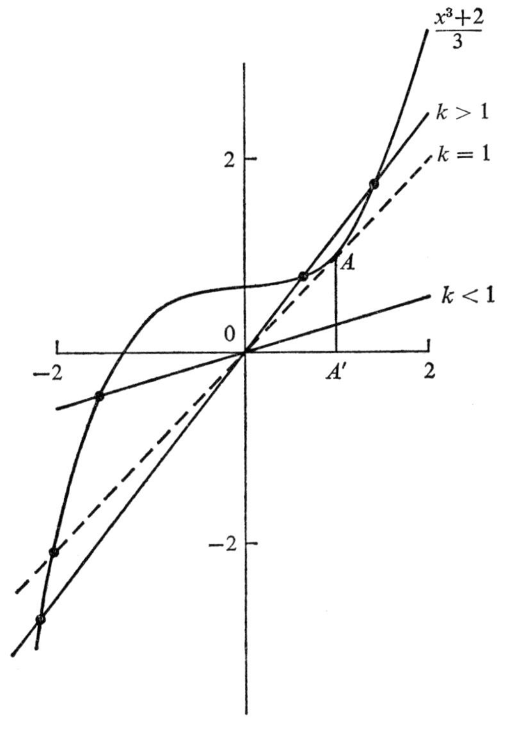

Fig. 5

We note that the graphs always cut at least once, so that there is always one root. But we note also a critical value of k where the graph of kx (shown by the broken line) *touches* that of $(x^3+2)/3$. If k is greater than this value, there are three roots. If it equals this value, there are two roots. If it is less than this value (including zero and negative k) there is only one root.

It remains to find the critical value.

Let us suppose the graphs touch where $x = a$. Then the length AA' is $(a^2+2)/3$ and hence the gradient of OA is $(a^3+2)/3a$.

But we can also find the gradient by differentiating $(x^3+2)/3$. From this it is found to be a^2.

Hence $(a^3+2)/3a = a^2$, or $a^3+2 = 3a^3$, or $2 = 2a^3$.

Hence $a = 1$, and $k = a^2 = 1$ (since k is the gradient of OA).

It follows that the equation has 3, 2 or 1 root according as k is greater than, equal to, or less than 1.

Exercise B

1. Sketch the graph of x^2-8x, and hence determine for what range of values of c $x^2-8x-c = 0$ has roots.

2. Sketch the graph of $x+36/x$, and hence determine for what range of values of b, $x^2-bx+36 = 0$ has no roots.

3. Repeat Question 2 by the method of Example 2.

4. Discuss the solubility of $x^3-3kx+2 = 0$ by sketching the graph of $(x^3+2)/x$.

5. Use the method of Example 2 to discuss the solubility of $x^4-kx+48 = 0$ for the different ranges of values of k.

6. (*a*) Find the turning value of $f(x) = ax^2+bx+c$ (where $a \neq 0$) and hence express the criteria for 2, 1 or 0 roots to the equation $ax^2+bx+c = 0$ in terms of a, b and c (separate the cases where $a > 0$ and $a < 0$).

(*b*) Prove that the following quadratic polynomials cannot be zero and state in each case whether it is always positive or always negative:

(i) x^2+5x+7,

(ii) $-1+3x-3x^2$,

(iii) $-12+7x-2x^2$,

(iv) $5-5x+2x^2$.

2. PHYSICAL PROBLEMS

There exist a fair number of problems in which, from considerations of simple geometry or probability, we can express one variable as a function of another and so come to obviously striking conclusions.

2.1 Simple problems.

Example 1. A square sheet of cardboard, 10 cm × 10 cm, has four equal squares cut out of the corners, so that the points A, B, C, D are the innermost corners of these squares. The cardboard is then folded along the lines

AB, BC, CD and DA, in such a way as to form a box without a lid. What is the greatest volume that the box can have?

Here it is evident that if the sides of the small squares are x cm we must consider a value of x between two extremes: $x = 0$, in which case the volume of the box is 0, since it has no height; and $x = 5$, where again the volume is 0, since although the height of the box is 5 cm, it is left with no base. Somewhere between these limits we get a maximum volume, and it is clear from these considerations alone that the graph of the volume, expressed as a function of x, must take very roughly the form shown in Figure 6 by the dotted line.

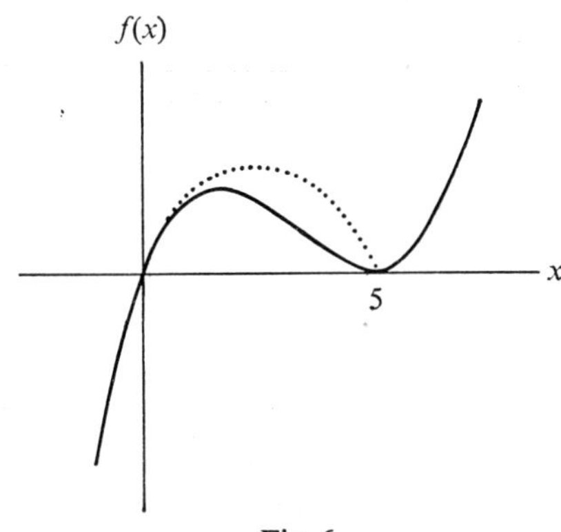

Fig. 6

More precisely, we can say that the height of the box is x cm; the side of the square base is $(10-2x)$ cm. Hence the volume is

$$x(10-2x)^2 \text{ cm}^3 = 4x(5-x)^2 \text{ cm}^3.$$

If we define $f(x) = 4x(5-x)^2$, we see that it has the graph shown in Figure 6; we verify that $f(0) = f(5) = 0$, and that there is a maximum somewhere between 0 and 5.

To calculate this exactly, we note that

$$f(x) = 4x(25-10x+x^2)$$

$$= 100x-40x^2+4x^3,$$

so that $$f'(x) = 100-80x+12x^2$$

$$= 4(5-x)(5-3x) = 0 \quad \text{where} \quad x = 5 \quad \text{or} \quad 1\tfrac{2}{3}.$$

We conclude that the greatest volume occurs where the side of the small square is $1\frac{2}{3}$ cm, and that the maximum volume is $2000/27$ cm^3.

Example 2. A sculler exercising himself sculls 500 m upstream at 3 m/s relative to the water. He then returns to his starting point at 1 m/s relative to the water. How long does he take?

The answer, evidently enough, is 'it depends on the rate of flow of the stream'. But if we suppose the rate of flow of the stream is constant (an unreasonable but characteristic assumption, when we are attempting a first approximation to the solution of a physical problem) then clearly we can express his time as a function of this rate of flow.

Let the rate of flow, or speed of the stream be x m/s. Then he travels upstream at a speed $(3-x)$ m/s relative to the bank; and returns at a speed of $(1+x)$ m/s. If his time is $f(x)$ seconds, then, since he moves 500 m in either direction,

$$f(x) = \frac{500}{3-x} + \frac{500}{1+x}.$$

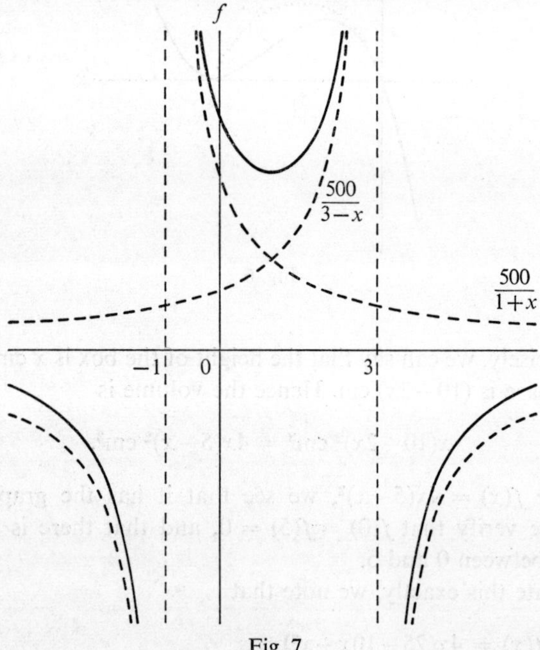

Fig. 7

Since we are given that initially he rows upstream, the domain of this function is restricted to positive x. It is however illuminating to sketch the graph for all values of x. We do it by adding the two graphs shown by the broken lines in Figure 7. Evidently then his time for $x = 0$ (that is, in still water) is 666·7 seconds. Thereafter the time passes through a minimum and tends to ∞ as x tends to 3.

It is easy to see why. If the speed of the stream is just less than 3 m/s, during the first part of his trip he only moves very slowly upstream and consequently takes a long time to reach his turning point.

If x is greater than 3, *the time he takes is negative.* This is not so absurd as it sounds. If the stream moves faster than 3 m/s, he will not, in the first part of his trip move upstream at all. He will move downstream and will never reach the point 500 m upstream. But, while he will not reach it, we can imagine him having come from it (having started some time before); and so, with the ordinary mathematical sign conventions we can think of him reaching it in a negative time.

It remains to calculate his minimum time, or what is from his point of view the 'best' speed of the river, if he is intent on using as little energy as possible.

We cannot differentiate $f(x)$ at present. But if we write it

$$f(x) = \frac{2000}{(3-x)(1+x)},$$

we see that it must have a minimum where $(3-x)(1+x)$ has a maximum. Let

$$g(x) = (3-x)(1+x) = 3+2x-x^2,$$

which is maximum where

$$g'(x) = 2-2x = 0 \quad \text{or where} \quad x = 1.$$

Hence the 'best' speed of the stream will be 1 m/s, giving a minimum time of 500 seconds.

The reader should try interpreting for himself the form of the graph if x is negative.

Exercise C

1. A man sculls upstream for 5 min at x m/s relative to the water, and then allows himself to drift back to his starting point with the stream which moves at 1 m/s. Express his average speed over the whole trip as a function of x, and show that it can approach, but not exceed, 2 m/s. Can it reach 2 m/s?

2. (i) A rectangle has perimeter 20 cm. Find the lengths of the sides when the area is a maximum.

(ii) A rectangle has area 36 cm^2. Find the lengths of the sides when the perimeter is a minimum.

3. A farmer wants to enclose three sides of a rectangular silo with 100 m of electric fencing, the fourth side being along a hedge. If the area is a function of the length of hedge used, discuss its nature in general terms, specifying zeros and domain. Find also the range.

4. What is the maximum area of a triangle whose sides are 10, 10 and $2x$ cm? (Consider the square of the area as a function of x.)

5. A length of wire 12 cm long is cut into two parts. One is formed into a square and the other into a rectangle whose length is twice its breadth. Find the greatest total area by expressing the area as a function of the breadth of the rectangle.

6. The size of a parcel dispatched through the post is limited by the fact that the sum of its length and the perimeter of its cross-section must not exceed 2 m. What is the volume of the largest parcel of square section which may be accepted for posting? (Express the volume as a function of the side of the square.)

7. A pack has twice as many red as black cards. Express as a function of the number of black cards the probability of drawing one black and one red card if two cards are drawn at random. Hence show the larger the pack the smaller this chance.

Show also that the chance of drawing two black and one red if three cards are drawn at random increases the larger the pack.

8. What is the greatest probability of having two or three dry days out of four? (Express the probability of having either two or three dry days as a function of the probability that any particular day is dry.)

9. A tea merchant spends equal sums of money buying two types of tea, the second K times more expensive than the first. He mixes them and sells them at a price between the two so that the difference between the selling price and cost price of the first tea is twice that of the second. For what range of values of K does he make a profit? (Express his profit on an outlay of £2 as a function of K.)

***2.2 Mathematical clothing.** The examples of the last section were all of an artificial simplicity. So too will be many of those in the present; but at least now we shall start to consider those techniques that mathematicians actually use when dealing with the physical world.

Consider the last exercise. In each question the reader was given a quite definite *problem*. He was to find something—a greatest value, for instance. Or he had to prove that one quantity decreased while another increased. Moreover he had to prove it about something also very definite: the probability of drawing a red card and a black one, or the time spent sculling on a river.

Now applied mathematicians rarely meet this kind of problem in practice. They come upon something which is much more vaguely and qualitatively stated. They have to make assumptions, in order to simplify the problem and make it mathematically manageable. They have to choose their own variables, and investigate the connections between them. And so all the way through the problem, until they come back to their original assumptions and consider—in the light of what they have done—in what way they could be improved in order to give a truer picture.

It is unlikely that the inexperienced reader will grasp much of this when stated (as above) in general terms. But it may be worth summarizing here a general drill useful when tackling problems, under numbered headings, so that he can trace it in the worked examples that follow.

When faced with a problem rather vaguely stated:

(i) Make initial assumptions to simplify the problem, and make it mathematically manageable.

(ii) Choose a dependent variable which will give an answer to the problem. (Often this is clearly implied in the problem itself.)

(iii) Make a list of the variables on which it depends.

(iv) Investigate the connections between these variables. (Often—almost always at this stage—we will find that, having chosen one of them, we can express all others in terms of it. If this is not so, it may be because our original assumptions were not sweeping enough; or because we have failed to make use of some of the conditions of the problem. Such a condition might state, for instance, that the volume of a variable solid is constant. In this case, give a name to the constant, and express the condition as a mathematical formula.)

(v) Choose an independent variable, express all other variables in terms of it, so that the original dependent variable (see (ii)) can be expressed as a function of it. (Make sure that the expression defining this function contains none but the independent variable and constants.)

(vi) Note the domain of the function, from the conditions of the problem.

(vii) Analyse the function, sketching its graph, and noting conclusions.

(viii) Return to the original assumptions and see whether they can be improved to give a more realistic picture of the problem. (This, though of much importance in more advanced work, will not really concern us here.)

All this may be described as giving a problem only vaguely stated (as in Example 1) a decent mathematical clothing. It has become common to describe it all—or the outcome of the initial stages—as a 'mathematical model'.

Example 1. How much metal is needed to make a tomato juice tin?

In this—as in many problems that float into the mathematicians world from those of commerce, or science, or administration—it is not at first clear what exactly is being asked for. One's immediate instinct is to reply with a number of rather facile 'it-depends-on.s': 'It depends on the amount of tomato juice; on the thickness of the metal; on the shape and proportions of the tin.'

It is an instinct to be resisted. After all, it is not hard to guess the essential nature of the problem. An industrialist has the job of imprisoning a certain quantity of juice within a container of a type which his customers are more or less used to buying. He may well want to know—the question suggests he does—the most economical use of metal by which he can achieve this.

(i) Therefore we will assume for a start that the tin is of normal shape;

that its volume is constant; and that we wish to find those proportions which will require least metal.

These are assumptions defining the nature of the problem. We turn now to those that will simplify it and make it manageable. Our first job here is to iron out small but distracting irregularities: such as the undulations in the circular ends, or the lip for opening the tin.

We will assume therefore that the tin is a perfect cylinder, completely full of juice.

We have also to tackle the question of the thickness of the metal. This (like so much else) is unstated. But let us assume (which may be true) that there is a standard thickness for tins in our class; that it does not depend on the proportions of the tin, but is a constant for the problem. Let us assume further that it is so small in comparison with the other quantities, that the volume of metal may be taken as proportional to the surface area. ('Volume equals surface area multiplied by thickness' gives, of course, an overestimate because of the curved surface area, and discrepancies round the inside of the rim. We are assuming the error is negligible.)

(ii) These initial assumptions—typical of a first and admittedly crude 'model'—lead us immediately to choose *surface area* as our dependent variable; since the mass of metal is proportional to the surface area. Let us call it A units of area.

(iii) The surface area, A units, will depend on the radius of the tin (let us say r units of length) and the height (h units).

(iv) These are not independent of each other. We are investigating variously proportioned tins with constant volume. Let us call it V units of volume. V will be constant in the work, just as A, r and h are variables. Giving the condition mathematical form, we see that

$$V = \pi r^2 h$$

so that we can express r in terms of h, or h in terms of r.

(v) Whether we choose r or h as our independent variable may not seem to matter very greatly. Both will show adequately how the surface area changes as we change the proportions of the tin. Here the only real preference is found to lie—as often—with mathematical simplicity. We know that

$$A = 2\pi rh + 2\pi r^2.$$

Looking at the formula for V, we see that if we express r in terms of h, we shall get a square root in the formula for A; but that if we express h in terms of r we shall not. Let us then choose r as the independent variable so that

$$h = V/\pi r^2,$$

and

$$A = 2\pi r \frac{V}{\pi r^2} + 2\pi r^2 = 2V/r + 2\pi r^2.$$

This gives us A as a function of r since both π and V are constants.

(vi) r can only be positive, obviously; but there is no other restriction on its value, since we can have a tin with very small radius and very large height, or one with very large radius and very small height.

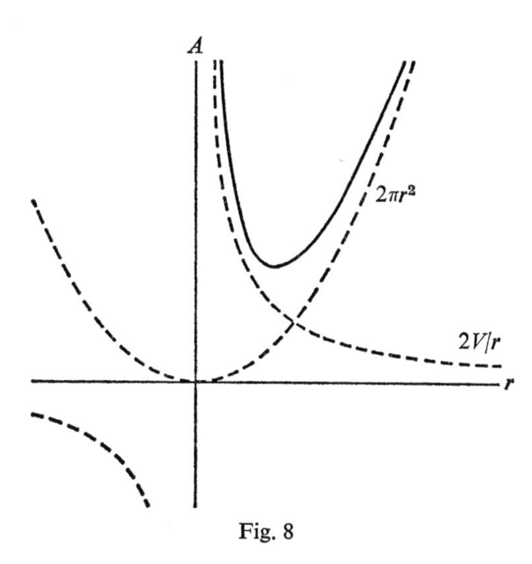

Fig. 8

(vii) Figure 8 shows the graph of A. $A \to \infty$ as $r \to 0$, and $A \to \infty$ as $r \to \infty$. Clearly then A passes through a minimum for some finite r. This, from the point of view of economy, will give us the best tin. Let us therefore calculate the minimum.

Put
$$f(r) = 2V/r + 2\pi r^2.$$
Then
$$f'(r) = -2V/r^2 + 4\pi r = 0$$
if
$$r^3 = V/2\pi.$$

At this stage it sometimes simplifies the algebra to give new names to such constants as V. Let us suppose for instance that $V = 2\pi$ units of volume. There is no loss of generality in doing this, since whatever the volume of the tin, we can choose a unit of volume (and hence of length) such that the tin has just 2π of them.

If we do so the minimum occurs where $r = 1$.

Since
$$V = \pi r^2 h = 2\pi,$$
this gives us $h = 2$.

Here then are the proportions of our 'ideal' tin: its height is just twice its base radius, or equal to its diameter. Many tins of course will not have these proportions; but that is because other considerations than economy (for instance appearance, habit, storage convenience and so on) guide the activities of industrialists. Still, with economy as our sole criterion, and within the limits of our 'model', this is the ideal tin.

(viii) We shall not spend much time trying to improve models in this book. But—to show our's was not the only manageable model—let us now take into consideration the rim at either end. The height and thickness of this will not, perhaps, vary with the proportions of the tin, being designed for the convenience of tin openers. Let us assume as our 'model' that it is a simple extension of the basic cylinder, and of constant height, H units say, much smaller than the radius and height of a normally proportioned tin.

Then, with the same notation as before, we get:

$$V = \pi r^2(h-2H) \quad \text{and} \quad A = 2\pi r^2 + 2\pi rh.$$

Hence $\quad A = 2\pi r^2 + 2\pi r(V + 2\pi r^2 H)/\pi r^2 = 2\pi r^2 + 4\pi rH + 2V/r.$

Hence the minimum surface area occurs where

$$4\pi r + 4\pi H - 2V/r^2 = 0$$

or

$$r^3 + Hr^2 - V/2\pi = 0.$$

Putting $V = 2\pi$, as before, we have to solve $r^3 + Hr^2 - 1 = 0$. We already know an approximate solution to this, $r = 1$, from the previous model. Taking this, and using Newton's method, we get as a second approximation,

$$r = 1 - H/(3+2H) \approx 1 - H/3.$$

Here then is a first approximation to the modification necessary to our previous answer if we wish still to preserve the minimum surface area.

The reader may ask why, if we had scruples about neglecting the rim, we did not do all this the first time. There is no very formidable reason in this example; though even here the work done on our first 'model' made that on the second somewhat easier. We followed the same algebraic procedure; we knew what sort of answer to expect (and hence a first approximation to the answer); we knew that H would be small compared to 1, and so on. But in more complicated work, the first model not only makes subsequent work easier; it is often the essential step which makes it possible. We shall not illustrate this here.

Example 2. Figure 9 shows a mathematical model of a flower. At night it folds its petals along the dotted lines. Their tips, A, B, C, D, E, F all meet in a point. What can we say about its surface area?

(i) What is uncertain (and therefore variable) in the figure (apart of course from the absolute size of the flower) is the relative size of adjacent petals. Clearly alternate petals are of equal size; but while, for instance, AP and BP are equal (or their tips will not meet) there is no reason why PQ and PS should be equal. Let us assume then that ACE is a fixed

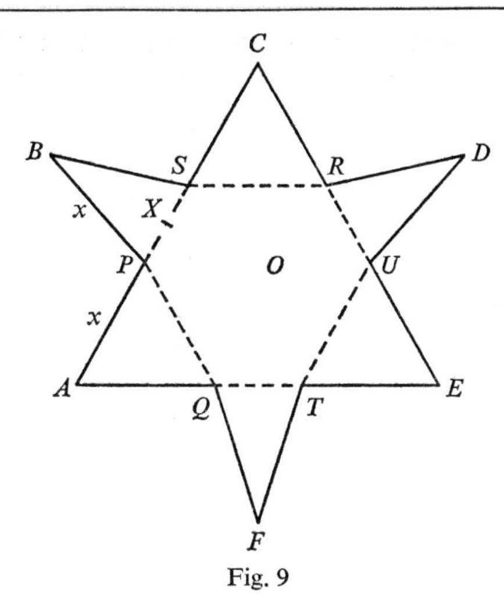

Fig. 9

equilateral triangle, but that all other points vary in such a way that, for instance, $AP = PB = BS = SC$, and $PSRUTQ$ vary on the sides of ACE.

(ii) The only elements of the surface area which vary in that case are the triangles BPS, DRU and FTQ. Since they are all equal, it is sufficient to take the area of one of them (BPS say) as dependent variable.

(iii) and (iv). The area of BPS depends on the lengths BP and PS. Both of these can be expressed in terms of the length AP (which we will call x units). For now $BP = x$ units; and $PS = AC-2x$ units, where AC is of constant length.

(v) We have already in effect chosen x as the independent variable. Let us choose the unit of length so that $AC = 2$ units. Then examining the triangle BPS, $PX = PS/2 = (2-2x)/2 = 1-x$. Thus applying Pythagoras's Theorem we see that

$$BX^2 = x^2-(1-x)^2 = 2x-1.$$

It follows that the area of BPS is $(1-x)\sqrt{(2x-1)}$. This suggests that, with our present knowledge, the correct dependent variable to choose is not the area, but the square of the area. Let us call this $f(x)$ units. Then

$$f(x) = (1-x)^2\,(2x-1).$$

(vi) Much of the interest of this example lies in the investigation of the domain. There are three geometrical conditions on x.

First, x must be less than (or in the limiting case equal to) half of AC. This is $x \leqslant 1$.

Secondly, x must be big enough for BPS to be a proper triangle. This means that $BP+BS > PS$; that is, $x+x > 2-2x$, or $4x > 2$, or $x > \frac{1}{2}$. In a limiting case again we can have $x = \frac{1}{2}$.

Finally the petals must be large enough to meet in a point at all. If they are too small, then clearly at night they will be left flapping vacantly at each other, unable to bridge the gap between them. In one critical case, where $x = 1$, they will meet, but the three petals like *BPS* will have been squeezed out of existence. In the other critical case, where they only just meet, they will all fold through 180° to exactly cover the central portion *PSRUTQ*, with their tips meeting at *O*. But *O*, the centre of *ACE*, is two thirds of the way down each altitude. Hence this will occur if *APQ* is (in linear dimensions) one third of *ACE*; that is, if $x = \frac{2}{3}$. Hence the third condition is $x \geqslant \frac{2}{3}$. (This swamps the earlier condition $x \geqslant \frac{1}{2}$.)

It follows that the domain is $\frac{2}{3} \leqslant x \leqslant 1$.

(vii) Thus fortified, let us return to $f(x)$. Figure 10 shows its graph, and the domain. It is worth noting that a hasty and injudicious glance *at the graph alone* might have led us to suppose that the domain was $\frac{1}{2} \leqslant x \leqslant 1$,

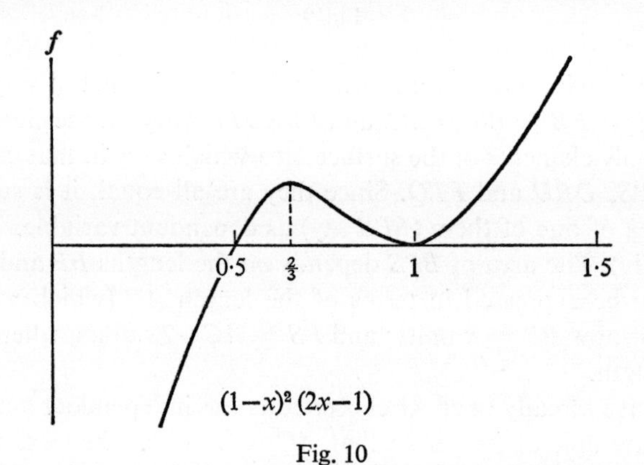

Fig. 10

which is untrue. It remains to decide whether the turning value *P* lies within the domain or (as shown in the figure) outside the domain, in which case the area of the flower decreases steadily as x increases from $\frac{2}{3}$ to 1.

Multiplying we get,

$$f(x) = 2x^3 - 5x^2 + 4x - 1,$$

so that

$$f'(x) = 6x^2 - 10x + 4.$$

But we already know that $f'(1) = 0$, so that $x-1$ is a factor of $f'(x)$. This helps us to find that

$$f'(x) = (x-1)(6x-4),$$

showing that we get a further turning value where $x = \frac{2}{3}$.

We conclude that the maximum occurs where $x = \frac{2}{3}$, that is, on the lower boundary of the domain, so that the area of the flower does steadily decrease as x increases from $\frac{2}{3}$ to 1. The maximum area thus occurs if the flower, at night, folds in such a way as to leave no obvious room for the pistils and stamen; but possibly that is because it is not a very good mathematical model.

We conclude with an example in which it is not very easy to choose a suitable dependent variable; and in which therefore we desert the order of presentation so far used in this section.

Example 3. Two men alternately toss a weighted coin. The second to toss is always allowed, by way of compensation, two tosses to the other's one. The first man to toss a head has won. For what weighting of the coin has the second man the greatest advantage over the first?

Here we evidently measure the weighting by the chance, p say, of throwing a head on any throw; this will be our independent variable. But what best measures the second man's advantage over the first is not at all clear. Let us therefore calculate the chance which each has of winning in any 'hand', or succession of three throws, as functions of p, and compare the two functions we get. The domain will be $0 \leqslant p \leqslant 1$.

The chance that the first man wins on the first three throws is simply the chance that he starts by tossing a head: that is, p. The chance that the second man wins on the three throws is the chance that the first does not throw a head, $(1-p)$, multiplied by the chance the second man throws a head either on the second throw or the third. These chances are p, and $(1-p)\,p$. Hence his total chance of winning on the first three throws is

$$(1-p)\,[p+p(1-p)] = p(1-p)\,(2-p).$$

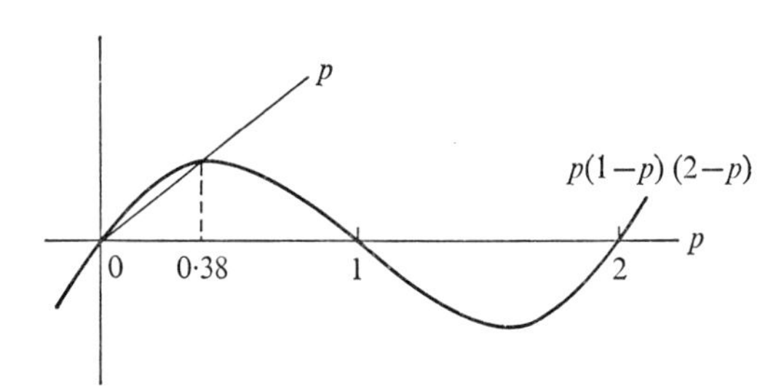

Fig. 11

We compare these functions by means of their graphs in Figure 11. Clearly the answer will depend on where they meet. This happens where

$$p = p(1-p)\,(2-p),$$

or

$$p^2-3p+1 = 0 \quad (p \neq 0),$$

or, in the relevant range, where

$$p \approx 0{\cdot}38.$$

This shows that the second man has a better chance of winning so long as the chance of a head is less than 0·38.

To decide what, from the point of view of the second man, is the best weighting we might proceed in a number of ways. We might calculate the value of p from which he has the greatest chance of winning. But this, disconcertingly, comes to about 0·42: that is, a weighting for which the first man has a greater chance still.

We might also consider the value of p for which the second man's chance of winning is the *greatest number of times* larger than the first man's. This is apparently more remunerative. If, for instance, they agreed to go on playing until one of them threw a head, it might seem to give the second man his greatest advantage. But the second man's chance of winning is in general $(1-p)(2-p)$ times greater than the first man's and in the domain $0 \leqslant p \leqslant 1$ (see Figure 12) this is greatest when $p = 0$. A surprising result at first sight. When $p = 0$, both their chances of winning are zero; a head will never be thrown, and the game will never come to an end.

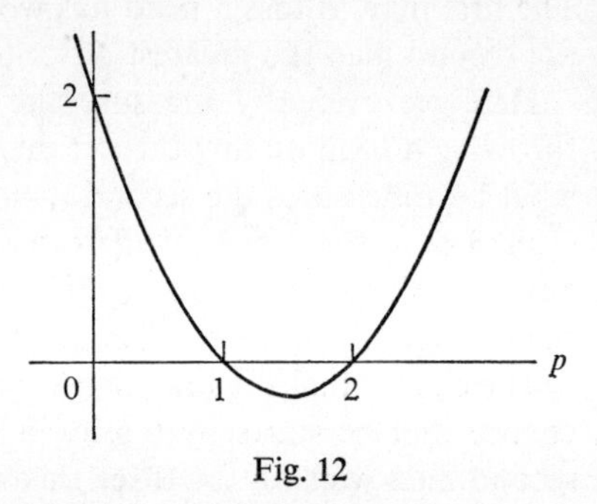

Fig. 12

We may conclude, possibly, that the second man's chance is, so to speak, the bigger zero; or more exactly, for small p, the second man's chance of winning is almost twice the first man's, though they would take correspondingly longer in reaching a decision at all; and it may be that he will think his advantage outweighed by the length of time it takes him to derive any benefit from it. Seen in this light, what he would regard as the best possible weighting becomes a matter of intimate personal psychology.

However we can get a definite answer if we suppose that the two men agree to play a definite number of 'hands'. For the first man will expect to win a fraction p of these; the second man a fraction $p(1-p)(2-p)$. The second man's advantage will depend on the amount by which $p(1-p)(2-p)$ *exceeds p*. It is easy to calculate that this is maximum if p is about 0·18.

Exercise D

1. A chocolate box 6 cm high is to have volume 1200 cm^3. It is tied round the middle one way with a single band of expensive ribbon and round the middle the other way with two of these bands; the bands lie in vertical planes. How should the manufacturer design the box if he wishes to spend as little as possible on ribbon? If the chocolates are about 2 cm wide, how would a girl wish the box to be designed in order to get as much ribbon as possible?

2. A rectangular tank is to be designed with a square base to hold a fixed quantity of water. Sketch the graph of its wetted surface area as a function of a suitable variable. What is the best design?

3. In order to satisfy the regulations of a cookery diploma, the pastry lining to the bottom and sides of a rectangular pork pie must be 0·5 cm thick. A girl has made 216 cm³ of pastry for this purpose and knows that the most satisfying pies are made with square bases. What is the biggest pie of this kind she can make?

4. A right circular cone fits tightly inside a sphere of radius 1 m. Discuss in general terms its volume as a function of the distance of the base from the centre. Specify the range and domain.

5. A nut of mass 1 kg is screwed at a distance of x cm from the centre of a rod of length 1 m and mass 4 kg. The whole is set oscillating about the centre of the rod in a vertical plane. It can be shown that the period of small oscillations is $2\pi\sqrt{\left(\frac{10000+3x^2}{2940x}\right)}$ seconds. What is the shortest possible period? (Consider $f(x) = (10000+3x^2)/x$.)

6. A store tent is to have volume 3 m³, and rectangular sloping faces closed by equilateral triangular ends. Can this be made from 10 m² of canvas? (Express the surface area as a function of the side of the triangle; and take $\sin 60°$ to be $\sqrt{3}/2$.)

7. A model of a pyramid on a regular 12-sided base is to be made by attaching lengths of wire together to form the 24 sides, and covering the triangular faces with cardboard. A boy has 10 m of wire for the purpose. Show that he does not need more than 0·91 m² of cardboard.

8. A rectangular doorway is to be built to fit beneath an arch of masonry shaped as an arc forming $\frac{1}{3}$rd part of a circle of radius 1 m whose centre is h m above the ground. Discuss, in general terms, the square of its area as a function of the height of the top of the doorway above the centre of the circle, (i) if $h = \frac{2}{3}$, and (ii) if $h = \frac{4}{3}$.

9. With the flower of Example 2, and another choice of variable, find how its volume varies.

10. A light circular hoop of radius 2 m is suspended from the ceiling by two 6 m strings which pass through a hole at the top of the hoop and whose other ends are attached to heavy equal rings threaded onto the hoop in the symmetrical manner shown in the figure. Assuming that there is a tendency in nature for heavy rings to assume the lowest possible position, discuss whether the hoop will tend to move up or down when its centre is $5\frac{1}{2}$ m below the ceiling.

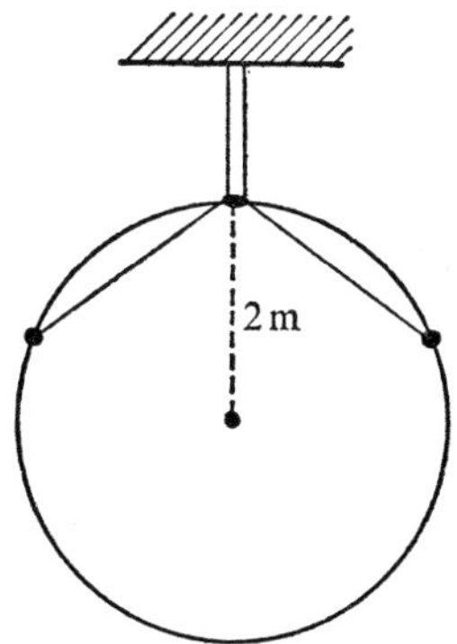

11. A closed cylindrical jam tin has a total surface area of 96π cm². By expressing the volume as a function of the radius, find whether a small increase in the radius will increase or decrease the volume when the radius is 3 cm. For what range of values of the radius is this true?

12. A rectangular enclosure is formed by placing end to end 96 hurdles. From the middle of one side a hurdle is removed to form a gate. How should the hurdles be arranged so that the distance from the middle of the gate to the far corner is minimum?

13. Two men *A* and *B* gamble, offering each other even chances on the result of tossing a weighted coin three times. If it comes down heads three times, *A* wins; if it comes down heads twice, *B* wins; otherwise nobody wins. Show that *B* has the better chance of winning so long as the chance of a head is less than $\frac{3}{4}$: and that by suitable weighting of the coin, *B* can make his chance of winning as many times greater than *A*'s as he likes. Why might there be little point in his doing this? If beforehand they agree to play a fixed number of games, what, from *B*'s point of view, is the best possible weighting?

14. The cross-section of a tunnel is an isosceles triangle of height $\frac{8}{3}$ m and base $2R$ m. Packing cases of height and width 1 m are to be pushed through it with their square ends parallel to the axis of the tunnel. By expressing the greatest length of such cases which (i) can be pushed through upright, and (ii) can be pushed through on their square ends, as functions of T, find for what values of R it is possible to slide a longer case by the first method than by the second.

15. Two playing cards of length 2 cm are leant one against the other, so as to form an upturned V, and so that a ping-pong ball can just be rolled between them. Show that, except for one particular case, by slightly altering the position of the cards, a bigger ping-pong ball can be rolled between them. (Express the square of the radius as a function of half the distance between the lowest points of the cards.)

5

RATE OF CHANGE

We shall now consider the use of derivatives in some physical problems. In these (see Chapter 1 Section 1.5) we can often express one variable as a function of another. How can we interpret the derivative in such cases?

1. RATE OF CHANGE

1.1 Rate of change. To clarify the nature of the problem we shall look at some examples, starting with the relation between the time of year and the length of daylight between sunrise and sunset. Figure 1 shows a graph of the duration of daylight throughout a year in latitude 50° North. It gives a general picture of the changes but is not sufficiently detailed to provide the figures which follow.

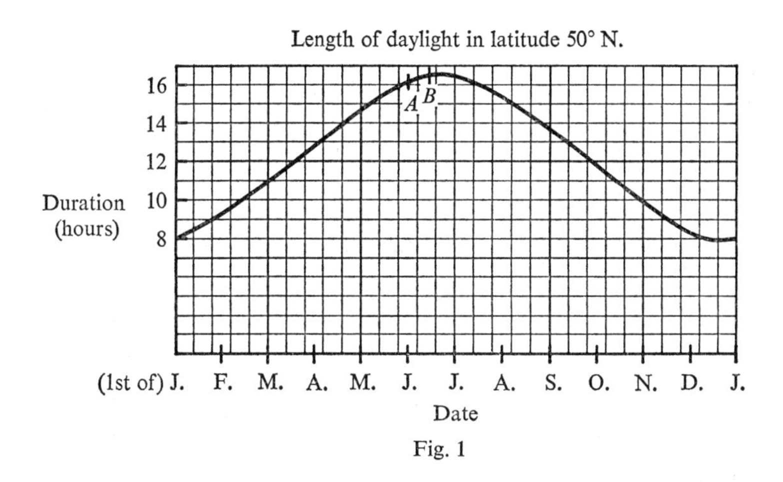

Fig. 1

In the fortnight 1 to 15 June the duration of daylight increases by about 17 minutes, that is at an average rate of about 1·2 min/day. But during this period the daily change is not constant, and if we are interested in the rate near 1 June we clearly get a better idea by taking the change from 1 to 2 June, which is in fact about 1·7 min/day. We cannot of course take an interval shorter than one day and we cannot talk about the rate of change of duration of daylight *on* 1 June, but only in an interval such as 1 June to 2 June.

Figure 2 is an enlargement of part of Figure 1, with the scale slightly distorted for convenience. BK represents the 17 min increase in the duration between 1 and 15 June; DE is the average of this for one day, that is, 1·2 min; while CE gives the actual increase of 1·7 min for the interval 1 to 2 June. It is often helpful to think of the average rate of change as being determined by the gradient of the chord: here AB, for the 14 day interval or AC, for the 1 day interval; since the gradient of AB, for instance, is the ratio BK/AK, which is 17 min/14 day or 1·2 min/day.

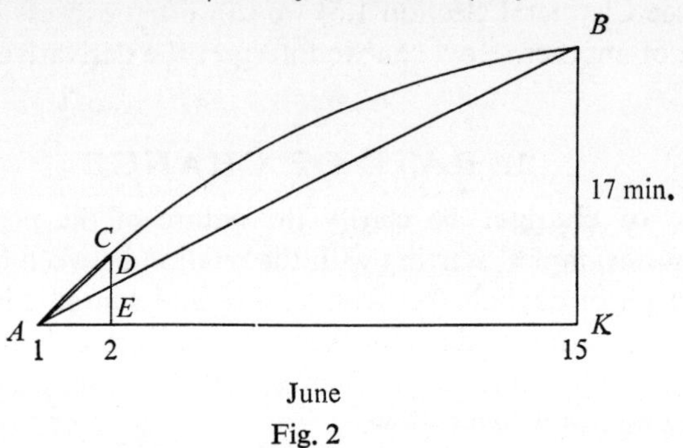

Fig. 2

In the next example the situation is different, as in this we can take intervals of any length, however small.

Figure 3 shows the graph of duration of daylight on 1 May in various northerly latitudes. The upward sweep of the graph shows that the duration

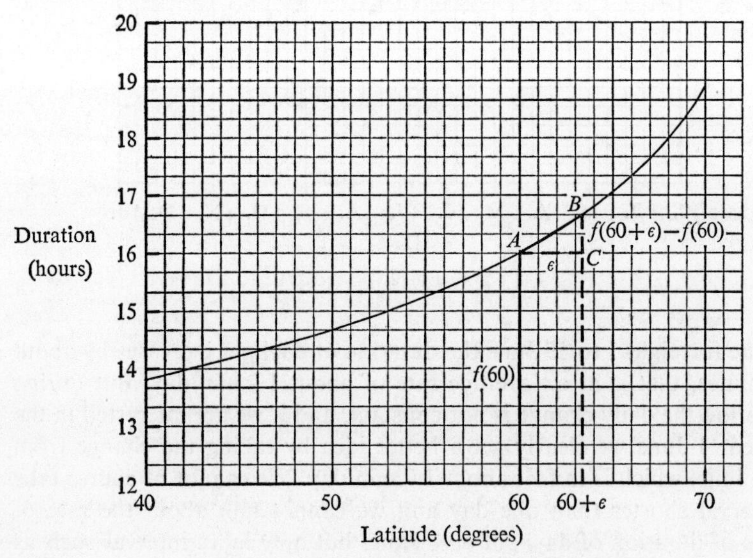

Fig. 3. Variation of daylight with latitude 1 May.

of daylight on that day increases with the latitude. Accurate values cannot be read from the graph, but in the appropriate tables it will be found that, from 60 to 70° N, the duration increases by 175 min; that is, by an average rate of 17·5 min/degree. Between 60° and 61°, the increase is found to be 12 min; that is, an average rate of 12 min/degree. We would get a different, but not very different, value for the average rate by taking intervals of less than one degree of latitude. Again, one can think of these average rates of change as being determined by the gradients of the chords corresponding to these intervals.

Let us now consider the matter algebraically, calling the latitude $L°$, and the duration of daylight D minutes, so that the graph is of a function f which maps L into D.

Now $f(60) = 958$, (from tables)

and
$$f(60+\epsilon) = 958+f'(60)\,\epsilon+s_\epsilon.$$

Between 60° N and $(60+\epsilon)°$ N, the increase in D is $f(60+\epsilon)-f(60)$ min, so that the average rate of change of duration with latitude is

$$\frac{f(60+\epsilon)-f(60)}{\epsilon}\ \text{min/degree}$$

over this interval of ϵ degree.

But
$$\frac{f(60+\epsilon)-f(60)}{\epsilon} = f'(60)+s_\epsilon/\epsilon,$$

where the value of s_ϵ/ϵ can be made as small as we like by sufficiently decreasing ϵ (we cannot annihilate ϵ as there would then be no interval over which to take the average).

This implies that we can make the average rate of change as near to $f'(60)$ min/degree as we like by taking a small enough interval. $f'(60)$ min/degree is an ideal value which the average rate of change can be made to approach as closely as we like (even though it cannot be attained) for small enough intervals; we shall call it the rate of change *at* 60° N.

It will be found that, to our accuracy of working, $f'(60) = 12$ and hence we can say that at 60° N the rate of change of the duration of daylight is 12 min/degree.

The geometrical interpretation of rate of change should be noticed. In Figure 3, $f(60+\epsilon)-f(60)$ is represented by BC while AC represents the change in latitude of $\epsilon°$. Hence the average rate of change over this interval is represented as we already know by the ratio BC/AC, which is the gradient of the chord AB. Now $f'(60)$ is the gradient of the tangent at A. So the gradient of the chord represents the average rate of change over an interval while the gradient of the tangent represents the rate of change at a point.

In general, for two physical quantities A, B, if A is a function of B such that $f(x)$ units of A correspond to x units of B, then we shall call $f'(p)$—in the appropriate units—the rate of change of A with B at the element p of the domain.

1.2 Rate of change for a linear function. A resistance wire 30 cm long carries a steady current. One end is at a potential of 39 V, the other at 45 V. The average rate of change of potential with distance along the wire is 6 V/30 cm or $\frac{1}{5}$ V/cm. Assuming the wire to be uniform, we may take the

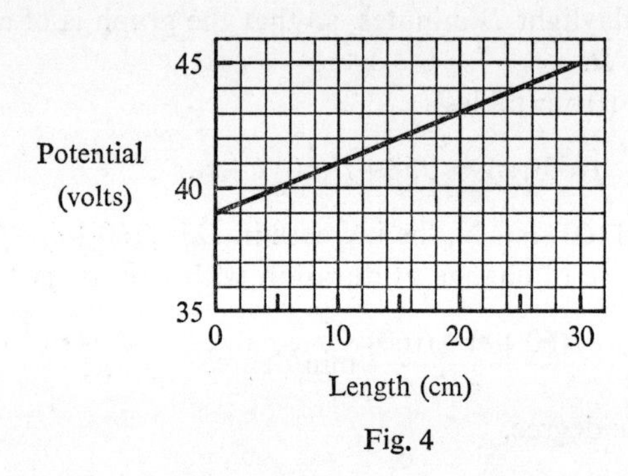

Fig. 4

potential at a point of the wire to be a linear function of its distance from its 'lower' end. In fact, if this distance is x cm, the potential is $39+\frac{1}{5}x$ V. Now the derivative of this is $\frac{1}{5}$ at all points of the wire, so we can say that the rate of change of potential with distance along the wire is $\frac{1}{5}$ V/cm at all points. In general we can see that the rate of change *at* a point of the domain is equal to the average rate of change over any interval of the domain whenever the function is linear.

1.3 Use of rate of change in approximations. We have noticed that a rate of change is often nearly the same as an average rate of change over a small interval. We can use this fact to make simple approximations. The method is illustrated in the following example.

Example. Suppose that when a mass of 3 kg is whirled round in a horizontal circle of radius 1 m at n rev/s, the tension T newton in the string is given by $T = 120n^2$. By approximately how much is the tension changed if n is increased from 2 to 2·01?

We could of course calculate the values of T when $n = 2$ and $n = 2{\cdot}01$ and take the difference, but instead we shall deal with the problem by evaluating the rate of change at $n = 2$.

If we write $f(n) = 120n^2$, so that $T = f(n)$, then $f'(n) = 240n$ and $f'(2) = 480$. That is, the rate of change of tension with speed of rotation is 480 N/rev/s when $n = 2$.

We now make the approximation that the average rate of change of tension with speed of rotation over the small interval from $n = 2$ to $n = 2{\cdot}01$ is very nearly 480 N/rev/s. So when the speed of rotation increases by 0·01 rev/s, the tension increases by about $480 \times 0{\cdot}01$ N, that is, about 4·8 N.

Sometimes it happens that the rate of change at a value of the domain, as it has been defined, bears a recognized name in applied mathematics. Such is the case in kinematics which we shall study in the next section.

Exercise A

1. Atmospheric pressure P N/m² as a function of height H metres is such that $P = f(H)$. In what units will the rate of change of pressure with height be given by the derivative?

2. On an income of £2000, a man would pay £250 tax, and if his income were £2500 he would pay £450. The average rate of change of tax with income is $\frac{2}{5}$ unit. What are the units?

3. On a fixed income a man would pay £250 tax if he had one child, but only £50 if he had five children. What is the average rate of change of tax with children?

4. The volume of a spherical balloon of radius r cm is $\frac{4}{3}\pi r^3$ cm³. What is the rate of change of volume with radius when $r = 2$? By approximately how much would the volume increase if the radius were changed from 2 cm to 2·001 cm?

5. The sag at the middle of a horizontal uniform beam supported at its ends is a function of the distance between the supports. If the sag is s cm when the distance is l metres, then for a particular beam $s = \frac{1}{8}l^4$. By approximately how much will the middle of the beam rise or fall if the ends are moved (*a*) out, (*b*) in, from a distance of 2 m by 10 cm?

6. Water is poured into a conical vessel. Suppose the volume of water in it, V cm³, as a function of the height, h cm, is given by $V = \frac{1}{2}h^3$. What is the rate of change of volume with height when $h = 10$? Approximately what increase of volume occurs when the height increases from 10 cm to 10·1 cm?

7. A circular ink blot spreads from an initial area of 1 cm² in such a way that if the radius is r cm after t seconds, $r = 3t+1$. Express the area, A cm², as a function of the radius and hence as a function of the time. What is the rate of change of radius with time? What is the rate of change of area with radius when $t = 1$? What is the rate of change of area with time when $t = 1$?

If the radius changes from 4 to 4·05 cm, find approximately, from a rate of change, the corresponding change in the area.

8. The volume of a cylinder, V cm³, is given in terms of the radius, r cm, and the height, h cm, by $V = \pi r^2 h$. If h has the fixed value 8, the volume is a function of the radius. What is then the rate of change of volume with radius when $r = 3$? What is the increase in the volume if the radius increases from 3 cm to 3·01 cm?

Alternatively, if r has the fixed value 3, so that the volume is a function of the height, what is the rate of change of volume with height? What is the increase in the volume when the height increases from 8 cm to 8·02 cm? Is your method approximate or exact?

If the radius increases from 3 cm to 3·01 cm, what approximate decrease in height from 8 cm will keep the volume constant?

2. KINEMATICS

Kinematics is the study of motion. It is a highly important subject in applied mathematics, but, apart from this, it is of interest at the present time for the illustration it provides of the practical meaning of a rate of change.

We shall confine ourselves here to the motion of a particle in a straight line.

2.1 Velocity. Figure 5 shows the graph of distance d metres plotted against time t seconds, for some motion of a particle moving in a straight line.

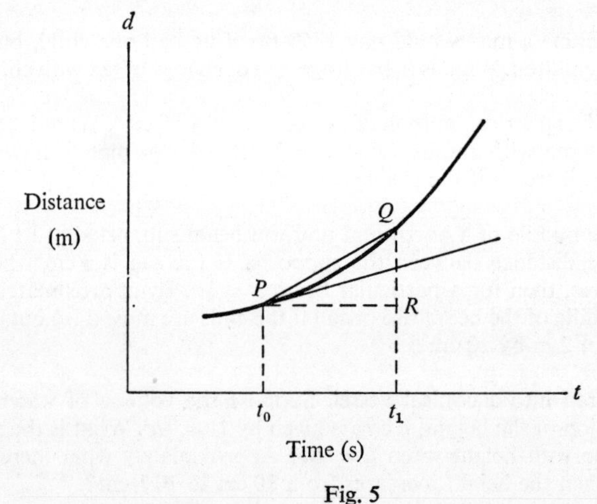

Fig. 5

In the time interval t_0 to t_1 seconds, represented by PR, the distance moved in metres is represented by QR, so that QR/PR m/s is the average velocity in this interval. This accords with our usual understanding of average velocity, but can be considered here as a definition of average velocity in terms of distance and time.

Now if f is the function which maps t into d, we know from our definition of rate of change that the rate of change of distance with time at t_0 is $f'(t_0)$ m/s. We also know that this is the value towards which the average velocity tends as the time interval is decreased. This rate of change is given the name velocity *at* time t_0, and it is represented by the gradient of the tangent to the graph at P.

Few people feel any difficulty with the idea of instantaneous velocity; our experience with speedometers makes us ready to accept its existence. However there are philosophical difficulties as we know from Zeno's paradox of the arrow, and indeed instantaneous velocity is quite a sophisticated notion. Essentially in applied mathematics it exists because of its definition (in terms of derivative).

In summary we can say that if a particle at time t units is a distance $f(t)$ units from some point of the line along which it is moving, then its velocity is $f'(t)$ at the instant t, the units being the ratio of those for distance and time.

Example. A particle is moving in a straight line in such a way that its distance from a point of its path is (t^2+1) metres at time t seconds after some zero time. What is its velocity at the instant when $t = 1$?

If s is the function which maps time into distance, then $s(t) = t^2+1$. The velocity at time t is therefore given by $s'(t) = 2t$, so that when $t = 1$ the velocity is 2 m/s.

This answers the question posed, but it may help to clarify ideas if we calculate the average velocities for some small time intervals from the instant $t = 1$ to compare with the velocity at $t = 1$.

To find the average velocity for the interval $t = 1$ to $t = 1{\cdot}1$, we first calculate $s(1)$ and $s(1{\cdot}1)$ and then the difference, which is the distance moved in that 0·1 second. Now $s(1) = 2$ and $s(1{\cdot}1) = (1{\cdot}1)^2+1 = 2{\cdot}21$, so that the distance moved is 0·21 m. Hence the average velocity in this interval of 0·1 second is 2·1 m/s.

Next we find in the same way the average velocity for the interval $t = 1$ to $t = 1{\cdot}01$. $s(1) = 2$ and $s(1{\cdot}01) = (1{\cdot}01)^2+1 = 2{\cdot}0202$, so that the average velocity in this interval of 0·01 second is 2·01 m/s.

The average velocity for the interval $t = 1$ to $t = 1{\cdot}001$, found in the same way, is 2·001 m/s; and we see that as the interval is shortened, the average velocity approaches 2 m/s, which is the velocity at $t = 1$.

2.2 Acceleration. Suppose a particle is moving along a straight line and its velocity is given as a function of the time by $v(t)$. Figure 6 shows a possible graph of $v(t)$ with some numerical values. When t increases from 3 to 7 the velocity increases from 10 to 30 m/s, so QR represents 20 m/s and PR represents 4 seconds.

We can describe this situation by saying that on the average the velocity has increased at the rate of 5 m/s² (or 5 (m/s)/s); it is common to describe this as the average acceleration over this interval, and we shall define average acceleration in this sense. It is, of course, the average rate of change of $v(t)$ over this interval. It is then natural to call the rate of change at time $t = 3$, the acceleration at this time. That is, in general, we define the acceleration at time t as $v'(t)$.

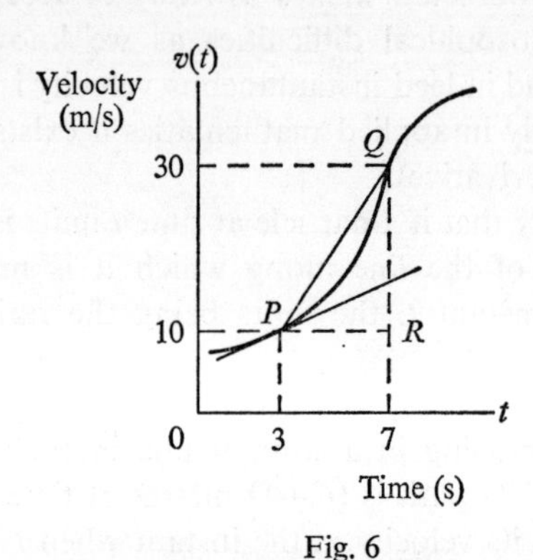

Fig. 6

As before we can note that the gradient of the chord PQ represents the average acceleration over the interval $t = 3$ to $t = 7$, while the gradient of the tangent at P represents the acceleration at $t = 3$.

To sum up, we can say that acceleration is the rate of change of velocity with time; and if velocity as a function of time is given by $v(t)$, then the acceleration is $v'(t)$.

2.3 Kinematical problems. In a particular problem of motion in a straight line, we are at liberty to choose both the origin for distance and the direction in which it is positive, as we please. We must also choose an origin for time; but then it is an established convention that later times are positive and earlier ones negative. The choice of sign conventions for distance and time determine those for velocity and acceleration; velocity being positive if distance is increasing. Thus, for example, a car losing reversing speed has a forward acceleration.

Example 1. A bead on a straight wire moves so that t seconds after passing through a fixed point O, its distance to the right of that point is $s(t)$ cm where $s(t) = t^3 - 11t^2 + 24t$. Determine the main features of the motion.

We can write

$$s(t) = t(t-3)(t-8)$$

which shows that the bead is at O when $t = 0$, 3 and 8. The graph of $s(t)$ plotted against t is shown in Figure 7(i). Now the gradient of the distance-time graph is the velocity $v(t)$ cm/s. Therefore

$$v(t) = 3t^2 - 22t + 24;$$

which can be written as

$$v(t) = (3t-4)(t-6)$$

so that $v(t) = 0$ when $t = 4/3$ or 6, and at these times $s(t) = 14{\cdot}8$ and -36. The velocity-time graph is shown in Figure 7(ii). Now the gradient of the velocity-time graph is the acceleration $a(t)$ cm/s². Therefore

$$a(t) = 6t - 22,$$

so that $a(t) = 0$ when $t = 3\frac{2}{3}$.

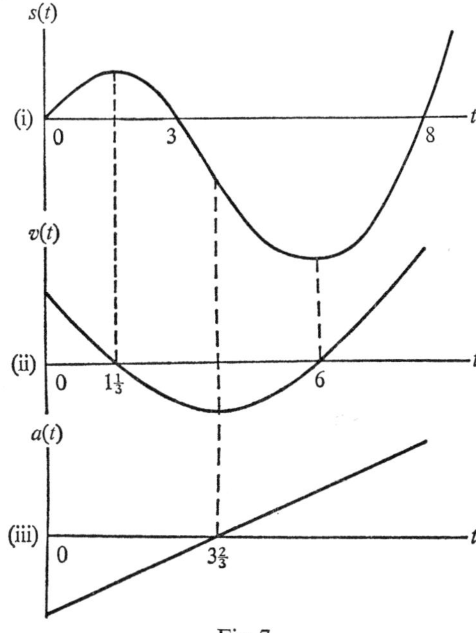

Fig. 7

The physical interpretation is that the bead moves to the right, reaching a maximum distance in this direction of 14·8 cm after $1\frac{1}{3}$ seconds. At that instant it is momentarily at rest. It then moves back towards O, passing through O at time 3 seconds; attains its greatest backwards velocity at time $3\frac{2}{3}$ seconds, after which it gradually slows down (accelerates forward) until it is once more momentarily at rest at time 6 seconds at a distance of 36 cm to the left of O. It then gradually gathers forward speed (since it continues to accelerate forwards) and passes through O for the last time at time 8 seconds, after which it moves away from O with ever increasing speed.

The special relationships existing between the three graphs are worth noticing, particularly that indicated by the middle broken line, where the greatest negative slope of the $s(t)$ graph corresponds to the lowest point of the $v(t)$ graph and to a zero of the $a(t)$ graph.

Example 2. A particle moves in a straight line so that its distance $s(t)$ cm from a fixed point of the line after t seconds is given by $s(t) = 8t^2 - t^4$. Find (i) when it is not moving, (ii) its initial speed and speed after 3 seconds, (iii) the distance moved and the average speed on the outward journey, (iv) when it returns to the starting point and at what speed, and (v) its acceleration when it is momentarily at rest.

Now
$$s(t) = 8t^2 - t^4 = t^2(8-t^2),$$
and its rate of change is the velocity $v(t)$ cm/s. Therefore
$$v(t) = 16t - 4t^3 = 4t(2-t)(2+t).$$
The rate of change of this function is the acceleration $a(t)$ cm/s². Therefore
$$a(t) = 16 - 12t^2.$$
If the particle is not moving, $v(t) = 0$, and this occurs when $t = 0$ or 2. It is therefore not moving initially and after 2 seconds. (i)

(Negative values of t have been disregarded, but they could be included, if required, provided the formula $s(t) = 8t^2 - t^4$ were known to hold for such values of t.)

The initial speed is given by the value of $v(0)$, which is zero; also since $v(3) = -60$, the speed after 3 seconds is 60 cm/s backwards. (ii)

(Moreover, as $s(3)$ is negative, it is in fact moving backwards: not towards the starting point but away from it.)

It moves outwards until it stops when $t = 2$, after which $v(t)$ becomes negative. As $s(2) = 16$ and $s(0) = 0$, the distance moved on the outward journey is 16 cm; and the average outward speed is 16 cm/2 s = 8 cm/s. (iii)

It returns when $s = 0$ and $t \neq 0$, that is when $t = \sqrt{8}$; and its speed then is $v(\sqrt{8})$, which is $-16\sqrt{8}$. Its speed therefore on returning to the starting point after $\sqrt{8}$ seconds is $16\sqrt{8}$ cm/s backwards. (iv)

It is momentarily at rest when $v(t) = 0$, that is when $t = 0$ or 2 as in (i); and its acceleration then is found by substituting in $a(t) = 16 - 12t^2$, giving the acceleration from rest initially as 16 cm/s² forwards, and after 2 seconds as 32 cm/s² backwards.

(The negative sign for $a(2)$ implies an acceleration towards the starting point since $s(2)$ is positive.)

Note on units. In this example, t, $s(t)$, $v(t)$, and $a(t)$ represent numbers, but the time is stated to be t seconds, and the distance $s(t)$ cm. If the time and distance had been stated in other units, the same motion would not have been given by $s(t) = 8t^2 - t^4$.

Exercise B

1. A particle moves on a straight line so that its distance $s(t)$ metres from a fixed point O of the line after t seconds is given by $s(t) = 10+9t^2-t^3$.

Find, (i) the average velocity for the first 2 seconds [use $s(t)$; note $s(0)$ is not zero];

(ii) the velocity after 4 seconds [find $v(t)$];

(iii) the acceleration after 1 second [find $a(t)$];

(iv) the velocity when the acceleration is zero.

2. A particle moves in a straight line so that t seconds after passing through a point O it is $s(t)$ cm from O, where $s(t) = 27t-t^3$.

Prove that it moves outwards from O for 3 seconds and then returns to O reaching it with a speed of 54 cm/s. Sketch the graphs of distance and velocity against time.

3. A particle moves on a straight line and is at a point O at zero time. After t seconds it is $s(t)$ cm to the right of O, where $s(t) = t^3-6t^2$. Determine the main features of the motion and draw graphs for distance, velocity, and acceleration, against time.

4. The velocity $v(t)$ m/s of a particle travelling in a straight line is given by $v(t) = 32-\frac{1}{2}t^2$ where t is the time in seconds. Find its initial velocity. Find at what instant the particle will reverse the direction of its motion, its acceleration then and the distance it has travelled from its starting point. Sketch the graphs of distance, velocity, and acceleration, against time.

5. If $s(t) = t^4$, where $s(t)$ represents the distance in metres and t the time in seconds, what are: (i) the velocity, (ii) the acceleration, (iii) the rate of change of the acceleration, when $t = 0$?

Is it possible for a body to be stationary and have no acceleration at a given time and yet be capable of immediate movement?

6. A spring whose upper end is fixed carries a mass at its lower end. If the mass is set in vertical motion, sketch the graph of the length of the spring as a function of the time. With the same axes and using a broken line, sketch the velocity-time graph, pointing out the chief correspondences between the two.

3. FUNCTION FROM DERIVATIVE

So far, simple rules have been discovered for finding the derived function of a polynomial, and use has been made of these derived functions in a variety of ways. Now we will investigate the reverse process of finding a function from a knowledge of its derivative.

3.1 Particular and general solutions. What is $f(x)$ if $f'(x) = x^2$? One solution is $f(x) = x^3/3$, but this is not the only one:

$$f(x) = \frac{x^3}{3}+3$$

also satisfies the equation. Two functions of x which differ by a constant have the same derivative; and hence the general solution of the equation $f'(x) = x^2$ is

$$f(x) = \frac{x^3}{3}+k,$$

where k is a constant. Particular solutions are

$$f(x) = \frac{x^3}{3}+7, \quad f(x) = \frac{x^3}{3}-2 \quad \text{and so on.}$$

Graphically, a change in the constant shifts the whole graph of $f(x)$ up or down and therefore does not alter its gradient for any value of x.

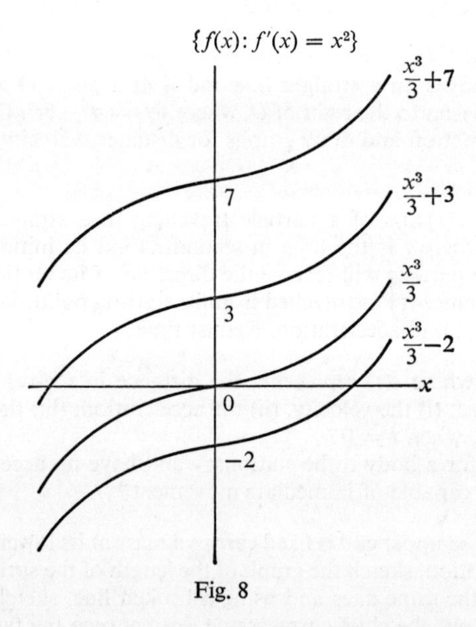

Fig. 8

In the 'general solution' the constant k is called an arbitrary constant; but if the value of $f(x)$ is known for any particular value of x, the value of k is then fixed and the solution becomes a 'particular solution'.

In this example, suppose we are given the additional information that $f(1) = 6$. The general solution is

$$f(x) = \frac{x^3}{3}+k,$$

but k must take a value such that $6 = \frac{1}{3}+k$, so the particular solution can only be

$$f(x) = \frac{x^3}{3}+5\tfrac{2}{3}.$$

The determination of the constant corresponds graphically to finding which of the solution curves passes through the point (1, 6). Information of the sort $f(1) = 6$ is commonly referred to as an initial or boundary condition. Something of its practical importance will be evident in the following examples.

Example 1. The velocity of a car in m/s, t seconds after 'zero' time, is given by $v(t) = 6t+2$. Find the distance in metres measured from some 'zero' position as a function of the time in seconds.

Let us denote the distance function by $s(t)$; then $v(t) = s'(t)$. We now have

$$s'(t) = 6t+2,$$

so that

$$s(t) = 3t^2+2t+c,$$

where c is a constant. This is a general solution.

If in addition we are told that distance is measured from the car's position when $t = 0$, so that $s(0) = 0$, then the particular solution is

$$s(t) = 3t^2+2t.$$

Example 2. A particle moving vertically under the influence of the earth's gravitational field, has a downward acceleration of about 9·8 m/s². If initially (when $t = 0$) it has an upward speed of 20 m/s, express its distance $s(t)$ metres, measured downwards from its initial position, in terms of the time t seconds.

Let $a(t)$ m/s² be the acceleration and $v(t)$ the velocity at time t seconds, where the positive direction is downwards for both. Then $a(t) = 9{\cdot}8$, and the initial conditions are

$$s(0) = 0 \quad \text{and} \quad v(0) = -20.$$

Now

$$a(t) = v'(t)$$

so that

$$v'(t) = 9{\cdot}8.$$

It follows that $v(t) = 9{\cdot}8t+k$, where from the initial conditions $k = -20$. Hence

$$v(t) = 9{\cdot}8t-20.$$

But

$$v(t) = s'(t),$$

so that

$$s'(t) = 9{\cdot}8t-20,$$

and therefore

$$s(t) = 4{\cdot}9t^2-20t+c,$$

where from the initial conditions $c = 0$. Hence.

$$s(t) = 4{\cdot}9t^2 - 20t.$$

Since the solution here was a two stage process, two sets of boundary conditions were necessary for finding a particular solution. The general solution involving two arbitrary constants is

$$s(t) = 4{\cdot}9t^2 + kt + c.$$

Example 3. Water is poured into an urn at a steadily increasing rate starting at zero and increasing at $\frac{1}{2}$ litre/s² for 5 seconds, and then at a steadily decreasing rate falling to zero after a further 3 seconds. How much water is poured into the urn altogether?

Let the rate of pouring be $r(t)$ litres/s, and the quantity of water in the urn be $q(t)$ litres after t seconds.

Then for $0 \leqslant t \leqslant 5$,

$$r'(t) = \tfrac{1}{2}.$$

So, $\quad r(t) = \frac{1}{2}t$ (zero constant, since $r(0) = 0$).

But the rate of pouring equals the rate of increase of the quantity of water in the urn. That is,

$$r(t) = q'(t).$$

Hence $\quad q'(t) = \frac{1}{2}t,$

which gives $\quad q(t) = \frac{1}{4}t^2$ (zero constant, since $q(0) = 0$). (i)

For $5 < t \leqslant 8$, the rate falls to zero when $t = 8$ from its value $\frac{5}{2}$ l/s (that is, $r(5)$) when $t = 5$. Hence

$$r'(t) = -\tfrac{5}{6}.$$

This gives $\quad r(t) = -\frac{5}{6}t + a$ (where a is a constant).

But $r(5) = \frac{5}{2}$, and therefore $\frac{5}{2} = -\frac{25}{6} + a$; so $a = \frac{20}{3}$. Replacing $r(t)$ by $q'(t)$, we get

$$q'(t) = -\tfrac{5}{6}t + \tfrac{20}{3};$$

from which $\quad q(t) = -\frac{5}{12}t^2 + \frac{20}{3}t + b$ (where b is a constant).

Now from (i), $q(5) = \frac{25}{4}$, so that $\frac{25}{4} = -\frac{125}{12} + \frac{100}{3} + b$. From this $b = -\frac{50}{3}$, and therefore

$$q(t) = -\tfrac{5}{12}t^2 + \tfrac{20}{3}t - \tfrac{50}{3}.$$

This gives as the quantity of water in the urn after 8 seconds:

$$q(8) = -\tfrac{5}{12}64 + \tfrac{20}{3}8 - \tfrac{50}{3} = 10 \text{ litres}.$$

Exercise C

1. What is $f(x)$ if $f'(x) = 2x$ and $f(0) = 1$?

2. What is $f(x)$ if $f'(x) = 10x^4 - 3x^2$ and $f(1) = 5$?

3. What is $f(x)$ if $f'(x) = 1 - \dfrac{2}{x^2}$ and $f(2) = 0$?

4. The speed of a car at time t sec is $1 + \frac{1}{12}t^2$ m/s. What is (i) its initial speed, (ii) its speed after 2 seconds, (iii) its acceleration after 2 seconds?

If the distance travelled from its initial position is $s(t)$ metres, express $s(t)$ in terms of t and find how far the car goes in the first 2 seconds. Hence find the average speed for the first 2 seconds. (This is not the same as the average of (i) and (ii).)

5. (i) The distance moved by a particle in a straight line from the origin, in time t seconds, is $t^3 + 3t - 4$ metres. Find the velocity and acceleration when $t = 5$.

(ii) The acceleration of a particle which starts from the origin at zero time, and moves in a straight line, is $3t - 2$ m/s², where t is the time in seconds. Find its velocity and its distance from the origin when $t = 4$.

6. A boy accelerates from rest at 0·8 m/s² for 4 seconds, after which his acceleration changes to $-0{\cdot}4$ m/s² without a sudden change of speed. How far does he go before stopping?

Sketch the graphs of acceleration, speed and distance plotted against time.

7. At time $t = 0$ a particle is moving at 30 cm/s, in a straight line, with an acceleration of $45 - 9t$ cm/s² until $t = 6$, where t is the time in seconds. When $t > 6$ its acceleration is zero. Find (i) its speed after 2 seconds, (ii) when its speed is greatest, and (iii) the distance travelled in the first 10 seconds.

Sketch graphs of acceleration, speed and distance plotted against time.

8. A forest fire at time t hours is spreading at the rate of $9t^2 + t^3$ m²/hour. What area is burnt out (i) in the first 4 hours, (ii) between $t = 2$ and $t = 10$?

[*Hint:* if the area burnt as a function of time is $A(t)$ m², what is $A'(t)$?]

9. The rate of charging the plates of a battery is $25 - t^2$ amps, after charging for t hours. How long will it take to give them a charge of 40 ampère-hour?

10. At an art gallery it is found that when the entrance fee is $5x$ p per person, the number admitted per day is a function of x, $f(x)$, where $f'(x) = 100(1-x)$ for $2 \leqslant x \leqslant 4$. When the fee is 10 p, 800 people are admitted per day. How many would be admitted per day if the fee were 20 p?

Sketch the graph of $f(x)$.

11. In a traffic analysis the rate of increase in the number of cars/minute streaming into the Svenndorffi capital in the early hours of the feast of St Venndorf is assumed to be a linear function of the air temperature T °C. So $f'(T) = a + bT$, where a, b are constants and $f(T)$ is the number of cars/minute as a function of T. It has been observed that the number of cars/minute at temperatures 4 °C, 6 °C, 10 °C were 130, 190, 250; what should be the number when the temperature rises to 20 °C?

12. If for a man in a particular locality the rate of absorption of radiation is given by k/t^2 (k is a constant, $t \geqslant 1$) where $t - 1$ is the number of days after an atomic explosion, find the ratio of the amount he absorbs in the first five days to that in the next five days.

***13.** The time indicated by a sand-glass is a function of the quantity of sand still to run out. If the time is $f(x)$ minutes when x cm^3 of sand are left, and then sand is running out at $5x^2/(1+10x^2)$ cm^3/minute, what is $f'(x)$? How long does it take for the volume to decrease from 4 cm^3 to 1 cm^3?

4. INTEGRATION

The process of finding the derivative of a function is commonly called *differentiation*, and we speak of *differentiating* x^3, for example, to obtain $3x^2$. The reverse process—with which we are preoccupied in this chapter—of finding a function from its derivative, is known as *integration*; and we can say that *integrating* $3x^2$ gives x^3+constant.

In the rest of this chapter we shall look at a variety of physical problems involving integration. The technique is the same in each case and relies on finding the linear approximation to a function f, where the physical quantity considered—mass, area, heat content or whatever it may be—is this function f of a variable such as length or time. This is done by finding how much the physical quantity changes for a small change in the variable. That is, from the information given we determine $\phi(x)$ in the expression

$$f(x+\epsilon) = f(x)+\phi(x)\,\epsilon+s_\epsilon.$$

We can then identify $\phi(x)$ as $f'(x)$, so that the general form for $f(x)$ can be found by integrating $\phi(x)$. If a particular value of the physical quantity is wanted at, for instance, $x = a$ it is $f(a)$.

4.1 Simple physical problems.

Example 1. If the width of a vertical cross-section of the Eiffel Tower at a depth of x metres from the top is $\dfrac{x^2}{1500}+10$ metres, find the area of this section if its total height is 300 metres.

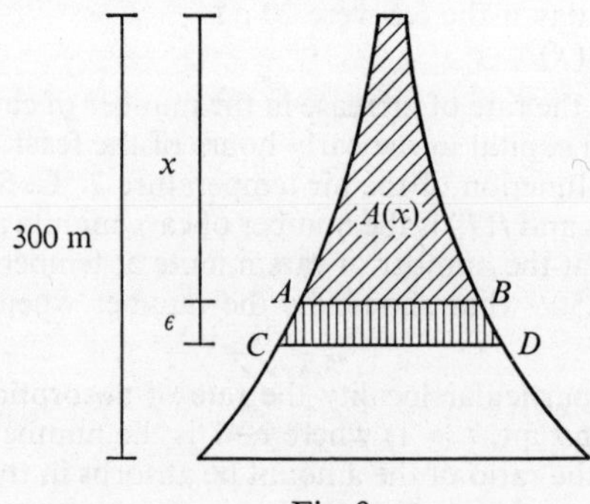

Fig. 9

Now the area from the top down to a depth of x metres is a function of x, which we shall denote by $A(x)$ m^2. Now $A(x+\epsilon)$ m^2 is the sum of $A(x)$ m^2 and the area of $ABCD$. The latter area is more than $AB\epsilon$ and less than $CD\epsilon$; but CD differs from AB by an amount which clearly tends to zero as ϵ tends to zero. The area is therefore $AB\epsilon+s_\epsilon$. Hence

$$A(x+\epsilon) = A(x)+\left(\frac{x^2}{1500}+10\right)\epsilon+s_\epsilon,$$

so that
$$A'(x) = \frac{x^2}{1500}+10.$$

Integrating,

$$A(x) = \frac{x^3}{4500}+10x \quad \text{(zero constant as } A(0) = 0).$$

The total area of cross-section is $A(300) = \dfrac{27\,000\,000}{4500}+3000 = 9000$ m^2.

Example 2. A spring of natural length 30 cm is stretched to 40 cm. How much work is done in stretching it, if the tension is $0{\cdot}2x$ N, when the extension is x cm? [For a constant force, work (Joule) = force (N) × distance (m).]

The work done in extending the spring x cm beyond its natural length is a function of x, which we denote by $W(x)$ Joules. Now consider the work done in extending the spring from an extension x cm to an extension $x+\epsilon$ cm; that is from A to B in Figure 10 When the length is OA the tension

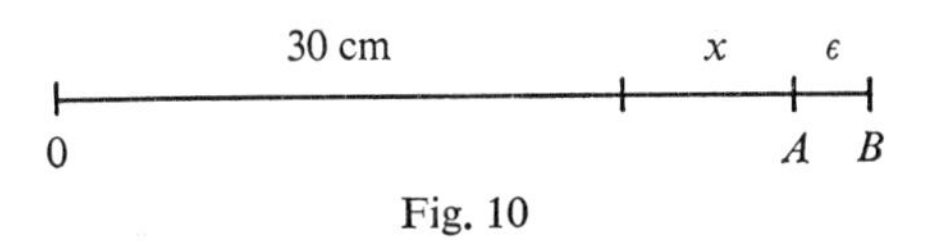

Fig. 10

is $0{\cdot}2x$ N, and when the length is OB the tension exceeds this by $0{\cdot}2\epsilon$ N. The work done in extending the spring slowly from A to B is the distance $0{\cdot}01\epsilon$ m multiplied by a force whose magnitude is somewhere between the tensions for the lengths OA and OB. It is therefore $0{\cdot}002x\epsilon+s_\epsilon$ Joules. Hence

$$W(x+\epsilon) = W(x)+0{\cdot}002x\epsilon+s_\epsilon,$$

so that
$$W'(x) = 0{\cdot}002x.$$

Integrating,

$$W(x) = 0{\cdot}001x^2 \quad \text{(zero constant, as } W(0) = 0).$$

The work done in stretching the spring 10 cm is $W(10) = 0{\cdot}1$ Joule.

Example 3. Find a formula for the volume of a right circular cone of base area A and height h. (A consistent system of units is assumed throughout.)

The volume of the top part cut off the cone by a plane parallel to the base at depth x below the vertex is a function of x, which we denote by $V(x)$.

The slice between this plane and a plane at a depth ϵ below it, is a frustrum between circular faces. The upper face, by the property of similar figures, has an area of $\frac{x^2}{h^2}A$, and the lower face differs from this by an amount which tends to zero as ϵ tends to zero. The volume of the frustrum is ϵ times an intermediate area; and therefore

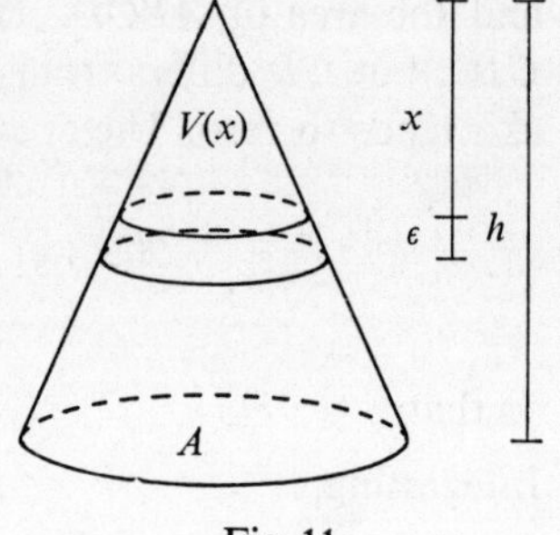

Fig. 11

$$V(x+\epsilon) = V(x)+\frac{x^2}{h^2}A\epsilon+s_\epsilon.$$

Hence $$V'(x) = \frac{x^2}{h^2}A.$$

Integrating this

$$V(x) = \frac{x^3}{3h^2}A \quad \text{(zero constant, as } V(0) = 0\text{)}.$$

The volume of the cone of height h is $V(h) = \frac{1}{3}Ah$.

The formula—a third of the area of the base times the height—holds for any cone or pyramid (not necessarily a 'right' figure) on any shaped base. The above proof is valid for all such cases.

Example 4. An iron poker shaped as a cylinder has cross-section area 0·5 cm² and length 30 cm. Its temperature decreases steadily from 200 °C at one end to 80 °C at the other. If it is allowed to cool to 30 °C along its whole length, how much heat is emitted?

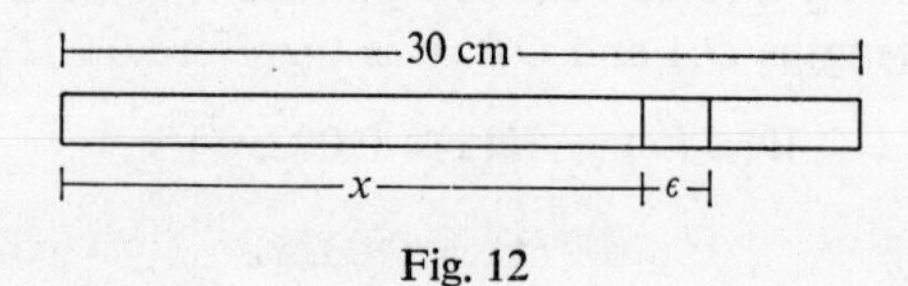

Fig. 12

If we measure distance from the hot end the temperature at distance x cm is $(200-4x)$ °C so that the temperature through which it cools there is $(170-4x)$ °C. At $x+\epsilon$ cm this is less by 4ϵ °C.

Now, tables of physical constants show that 1 cm³ of iron emits about 3·6 Joules of heat for each °C its temperature falls. Hence the heat emitted by the section from x to $x+\epsilon$ cm, of volume $0{\cdot}5\epsilon$ cm³, is

$$3{\cdot}6(170-4x)\,0{\cdot}5\epsilon+s_\epsilon.$$

Therefore, if the heat emitted from the x cm length of the poker, as a function of x, is denoted by $H(x)$ Joules, then

$$H(x+\epsilon) = H(x)+1{\cdot}8(170-4x)\,\epsilon+s_\epsilon.$$

Therefore $\quad H'(x) = 1{\cdot}8(170-4x).$

Integrating, $\quad H(x) = 1{\cdot}8(170x-2x^2)\quad$ (zero constant as $H(0) = 0$).

Hence the total heat loss for the poker is $H(30) = 5940$ Joules.

Example 5. Find the total mass of a bar of wood of 0·1 m² cross-section and length 3 m, where the density varies non-uniformly from 1500 kg/m³ at one end to 1800 kg/m³ at the other, according to the formula

$$1500+130x-10x^2 \text{ kg/m}^3$$

at x m from the light end.

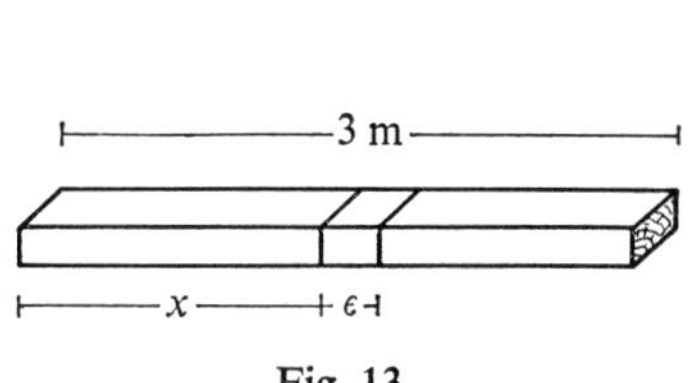

Fig. 13

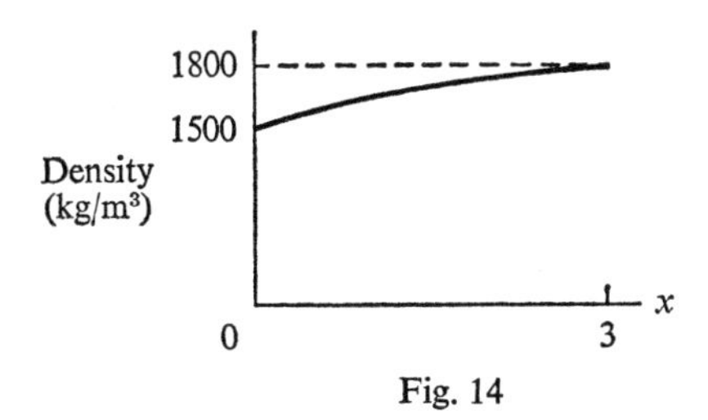

Fig. 14

The graph of the density against length along the bar is shown in Fig. 14.

The mass of the part of the bar up to a distance x m from the light end is a function of x which we denoted by $M(x)$ kg. The volume of the section whose ends are at distances x and $x+\epsilon$ m from the end is $0{\cdot}1\epsilon$ m³ and its density varies from $1500+130x-10x^2$ kg/m³ to a value which differs from this by an amount which tends to zero as ϵ tends to zero. Hence the mass of this section is $0{\cdot}1\epsilon(1500+130x-10x^2)+s_\epsilon$ kg.

Hence $\quad M(x+\epsilon) = M(x)+(150+13x-x^2)\,\epsilon+s_\epsilon,$

so that $\quad M'(x) = 150+13x-x^2.$

Integrating

$$M(x) = 150x+\tfrac{13}{2}x^2-\tfrac{1}{3}x^3 \quad \text{(zero constant, as } M(0) = 0).$$

The total mass of the bar is $M(3) = 499{\cdot}5$ kg

$$\approx 500 \text{ kg.}$$

This is not, of course, the same as if the density had been constant throughout at the mean value of 1650 kg/m³, when the total mass would have been 495 kg. However, in the last example where the temperature drop was uniform along the poker, we could instead have assumed that the temperature had the constant value of 110 °C above the surroundings.

In this example we have used the idea of density at a *point*. If we say the density at a point is 1500 kg/m³ we mean that for a small region enclosing that point, the average density (mass÷volume) is approximately 1500 kg/m³, and that it can be made as close as we like to 1500 kg/m³ by taking a small enough region.

Example 6. A thin horizontal metal plate shaped as shown in Fig. 15, where the curved edges are such that their width apart at a distance x m from the straight edge AB is $2+x-x^2$ m, is being lifted at the point T, 2 m from AB, so that it turns about AB. If the density of the plate is 600 kg/m², find the least force needed.

If the upward force is F N its moment about AB is $2F$ Nm. This must equal the sum of the moments about AB of the gravitational forces over the whole surface.

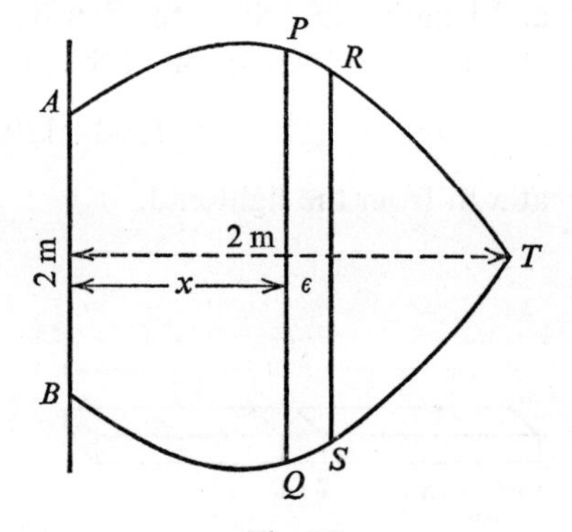

Fig. 15

Now the weight of the strip $PQSR$ whose parallel edges are x and $x+\epsilon$ m from AB is its area times the weight per unit area; that is

$$(2+x-x^2)\,\epsilon 600\,g+s_\epsilon \text{ N},$$

where $g = 9{\cdot}81$, since the width of PQ is $2+x-x^2$ m and the width of RS differs from this by an amount which tends to zero as ϵ tends to zero.

The moment of the weight of this strip is found by multiplying this weight by a leverage which lies somewhere between x and $x+\epsilon$ m. It is therefore $(2+x-x^2)\,\epsilon 600\,g\,x+S_\epsilon$ Nm. Now if we denote the moment of the weight of the part of the plate between AB and PQ, which is a function of x, by $M(x)$ Nm, we have

$$M(x+\epsilon) = M(x)+(1200x+600x^2-600x^3)\,g\,\epsilon+s_\epsilon,$$

so that
$$M'(x) = (1200x+600x^2-600x^3)\,g.$$

Integrating,

$$M(x) = (600x^2+200x^3-150x^4)\,g \quad (\text{zero constant as } M(0) = 0).$$

The total moment of the weight of the whole plate about AB is

$$M(2) = 1600g \text{ Nm}.$$

Hence $2F = 1600g$; so the required force is $800g$ N $\approx$ 7850 N.

Exercise D

1. The figure shows a map of a part of a river. Calculate its area if the width of the river parallel to the end section AB is $(12+10x-x^2)/10$ cm at a distance x cm from the line AB, and the total distance between the end sections is 9 cm.

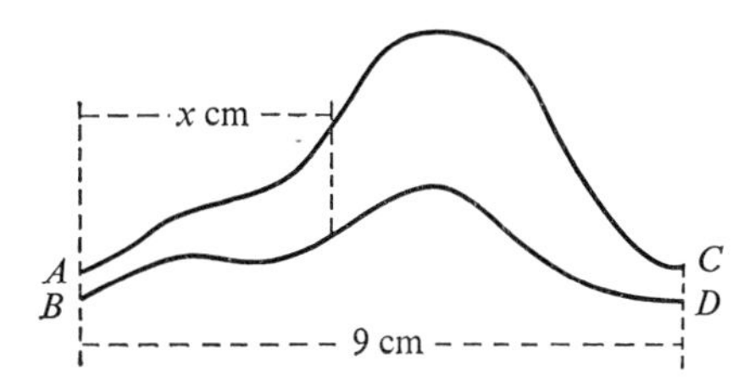

2. Calculate the volume of a cooling tower 50 m high which has a cross-section area of $(x^2-60x+2000)$ m² at a height of x metres.

3. The cost of laying a railway track in difficult country x miles from the base depot is £$(10000+40x+0{\cdot}1x^2)$ per mile. How much does it cost to lay a section 100 miles to 200 miles from the base depot?

4. The population density feeding an arterial road into London with commuter traffic is $320+24x$ persons per mile, where x miles is the distance from the London end of the road. If the cost of the journey is 5 p per mile, what is the total cost of the inbound traffic for all people using this road who live within 10 miles of its London end?

5. The water pressure at a point of a rectangular lock gate is kx N/m², where k is a constant and x metres is the depth below the water surface. What is the total force from the water pressure on the gate if the gate is a metres wide and its lower edge is b metres below the water surface?

6. The metal part of a gun barrel has a cross-section area of $(0{\cdot}04+0{\cdot}2x^{-2})$ m² at a distance of x metres from the breach. What is the volume of metal in the barrel if the barrel starts 3 m from the breach and is 4 m long?

7. In Example 6, the moment about AB of the force found is equal to the moment of the weight of the plate acting through the centre of gravity. Find the weight of the plate and hence the position of the centre of gravity.

8. The salinity of the ocean at a particular place is $(12+630x^{-2})$ kg/m³, where x metres is the depth below the surface for depths greater than 30 m, and 12·7 kg/m³ for lesser depths. What is the total amount of salt in a column of sea water there, of cross-section 1 m² reaching down from the surface to a depth of 210 m?

9. The density of a bar 2 m long of uniform cross-section 0·01 m² is $(400-x)$ kg/m³ at x metres from the end A. What is its mass in kg?

10. Calculate the sum of the moments about A of the gravitational forces on the bar in Question 9. Hence find the force required to support the bar at its mid-point, if the bar is freely hinged at A.

4.2 Further physical problems.

A few more examples of a slightly harder nature but depending on the same form of argument as before will complete the chapter.

First we shall derive the formula for the area of a circle using these methods. So far when a physical quantity has been expressed as a function of a variable, an increase in that variable has caused the physical quantity to 'spread' either along a straight line or at right angles to a straight line front. In this problem however, it is convenient to consider the area as a function of the radius and bounded by a circular front which spreads outwards concentrically.

Area of a circle. Let us denote the area of a circle of radius r by $A(r)$. It is reasonable to suppose that the area of the ring between circles of radius r and $r+\epsilon$ is greater than ϵ times the circumference of the inner circle and less than ϵ times the circumference of the outer one. Now the circumference of a circle of radius r is $2\pi r$ by the definition of π; so the area of the ring is between $2\pi r\epsilon$ and $2\pi(r+\epsilon)\,\epsilon$, and is therefore $2\pi r\epsilon + s_\epsilon$.

Hence $$A(r+\epsilon) = A(r)+2\pi r\epsilon+s_\epsilon,$$

so that $$A'(r) = 2\pi r.$$

Integrating, $$A(r) = \pi r^2 \quad \text{(zero constant as } A(0) = 0).$$

That is: the area of a circle of radius r is πr^2.

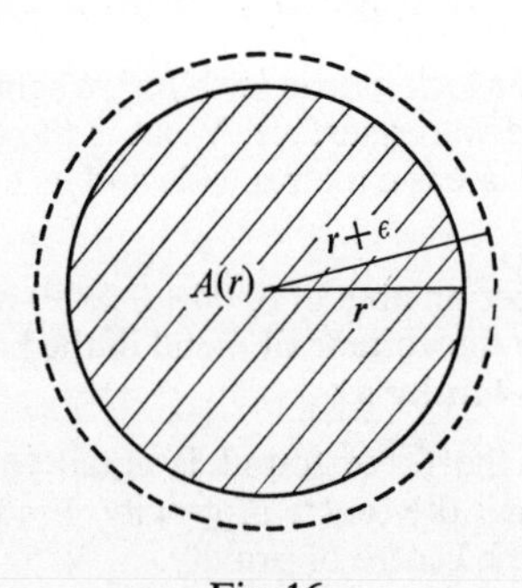

Fig. 16

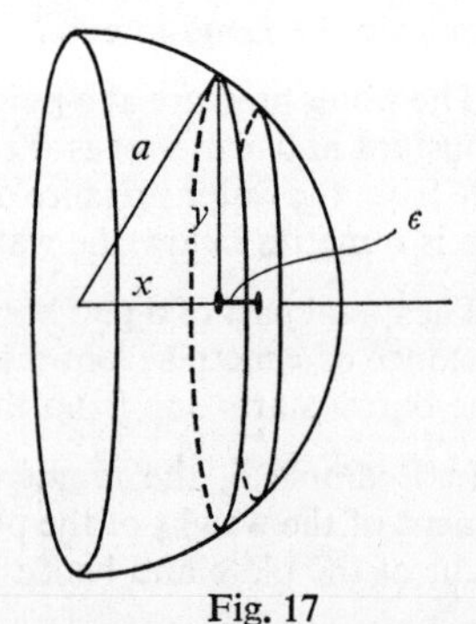

Fig. 17

Volume of a sphere. It is convenient algebraically to find first the volume of a hemisphere. For this we shall consider the volume lying between the plane face of the hemisphere and a parallel plane distance x away, as a function of x. The volume increases as the latter plane moves further from the diametral plane.

Figure 17 shows a thin slice of a hemisphere of radius a, between circular sections at a distance x and $x+\epsilon$ from the centre, parallel to the base of the hemisphere. The area of the section distance x from the centre is πy^2 where $y^2 = a^2-x^2$. The area of the other circular face differs from this by an amount which tends to zero as ϵ tends to zero. The volume of this slice, which is roughly frustrum shaped, is therefore $\pi y^2\epsilon+s_\epsilon$, that is $\pi(a^2-x^2)\,\epsilon+s_\epsilon$.

If we denote the volume between the base and the section distance x away by $V(x)$, then

$$V(x+\epsilon) = V(x)+\pi(a^2-x^2)\,\epsilon+s_\epsilon,$$

so that,
$$V'(x) = \pi(a^2-x^2).$$

Integrating,

$$V(x) = \pi(a^2x-x^3/3) \quad \text{(zero constant as } V(0) = 0).$$

The volume of the hemisphere is

$$V(a) = 2\pi a^3/3.$$

Hence the volume of the sphere of radius a is $4\pi a^3/3$.

Surface area of a sphere. We can use the formula for the volume of a sphere to find its surface area by considering the way in which the volume of a sphere of radius r increases as it expands outwards concentrically. For convenience we use Figure 16 again, but now pictured as representing the cross-section of a sphere of radius r whose radius is increasing.

Let the surface area of the sphere of radius r, as a function of r, be denoted by $A(r)$ and its volume by $V(r)$. Then the volume of the shell between concentric spheres of radius r and $r+\epsilon$ will have some value between $A(r)\,\epsilon$ and $A(r+\epsilon)\,\epsilon$, and as the difference between $A(r+\epsilon)$ and $A(r)$ tends to zero as ϵ tends to zero, we can say that

$$V(r+\epsilon) = V(r)+A(r)\,\epsilon+s_\epsilon.$$

Hence
$$V'(r) = A(r).$$

But
$$V(r) = 4\pi r^3/3 \quad \text{so that} \quad A(r) = 4\pi r^2.$$

The surface area of a sphere of radius r is therefore $4\pi r^2$.

Exercise E

1. In a city school the children are drawn from a region in which the population density is roughly $3x$ children/mile^2 at distance x miles from the school (in any direction) for $1 \leqslant x \leqslant 5$, and is zero for $x < 1$. The cost of travel to school is 2 p/mile. How many children live within 5 miles of the school; and what is the total cost for them to go to school each morning?

2. It is observed that 1 m from the centre of a spherical light-globe of radius $\frac{1}{2}$ m, moths are distributed, on average, at 20/m³. Assuming that in general the population density is inversely proportional to the fourth power of the distance from the centre, how many moths are there around the light? Within what distance from the centre are half the moths?

3. A uniform disc of radius a and weight W, is set spinning about its axis and then placed on a horizontal table. If the pressure from the table supporting the disc is constant over the surface and the frictional force over a small area is μx the normal contact force over that area, find the moment about the axis which is slowing down the disc.

4. The density of a dwarf star at distance r from the centre is $p+k(a-r)^2$, where p and k are constants and a is the star's radius. Find its total mass in terms of these constants.

5. A billiard ball is at the centre of a mathematical model of the universe in which stars are uniformly distributed throughout space. The quantity of illumination falling on the ball from stars distant between x and $x+\epsilon$ km away is proportional to the number of stars there and inversely proportional to x^2 (neglecting ϵ terms). Discuss the total quantity of illumination falling on the ball from stars up to X km away. Have you any major criticisms of the model?

6

COMPOSITE FUNCTIONS

So far we have learnt to differentiate only the most simple functions of x: the powers of x, their multiples, polynomials and x^{-1}. In this chapter we shall consider two types of composite function, which will enable us to use this knowledge in more complicated work.

1. COMBINING FUNCTIONS

1.1 Successive mappings. We think of a function as mapping one set of numbers onto another, and one of our ways of illustrating this has been to display the elements of the two sets on parallel lines, with arrows joining the elements of the domain to their images. We can now think of another function which maps the elements of the second set onto a third set, represented by a third parallel line.

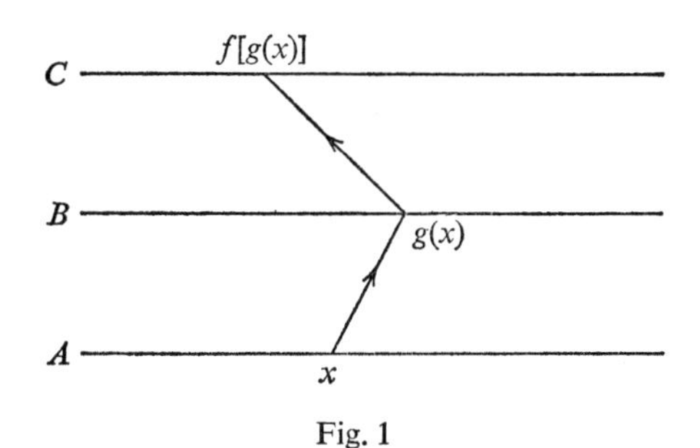

Fig. 1

In Figure 1, suppose g is a function which maps the set A onto the set B, and f is a function which maps B onto C. Then we can think of the variable x of A undergoing two successive mappings: in the first it maps onto $g(x)$ of set B, and from there onto $f[g(x)]$ of set C. Hence we have a mapping of the elements of A onto C, which defines a new function. Let us call this function h. A is the domain of h, while C is its range; and it is adequately defined by the statement:

$$h(x) = f[g(x)].$$

Now the function h is clearly defined if we know the functions f and g, and if the range of g is the same set as the domain of f. Hence we can speak of h as a *combination* (in a certain order) of the functions f and g.

In fact we shall speak of it as the *product* of f and g, denoted by the symbol fg. fg is defined by the equation:

$$fg(x) = f[g(x)].$$

The word product suggests an analogy with the multiplication of numbers, and the analogy is a close one, as can be seen if we are prepared to consider what numbers *are*. What is meant exactly by 'the number 3' is not an easy question to answer. We cannot easily define 3; we can only illustrate it. We cannot for instance say that it is $2+1$, for this is a mere subterfuge, begging the question of what is 2 and what is 1. But we can talk about 'three fingers', or 'three pieces of chalk', or 'three BMC motors', all of which are perfectly definite objects illustrating the same mathematical idea. '3' in fact, is (in vague terms) the mathematical abstraction lying behind all the collections of three objects we can think of: three fingers, three books and so on. '3' by itself means nothing in physical terms; '3 pieces of chalk' does. It is the perception of a common idea behind '3 pieces of chalk' and '3 BMC motors' which makes us mathematicians: an idea that was probably not recognized in the days when people evolved different words, such as 'pair', 'brace', 'couple' and so on, to denote twos of different things.

Now much the same can be said on the more abstract plane of function. Of course, in Chapter 1, we gave a precise definition of what a function was. But if we are familiar only with the world of numbers, the idea of a function may be rather perplexing. We cannot say exactly what the square function, for instance, P_2, *is*. $P_2(3)$ or $P_2(-5)$ means something in terms of numbers, just as '3 pieces of chalk' means something in terms of physical objects. P_2 by itself does not.

Just as we have to *operate* with the number 3 on some physical object, say an apple, to get '3 apples' and illustrate the meaning of 3 in the physical world, so we have to *operate* with P_2 on a number, say -5, to get 25 and illustrate its meaning in the world of numbers. We conclude that functions are mathematical abstractions with the same degree of abstractness *vis-à-vis* numbers, as numbers have *vis-à-vis* the physical world.

This analogy is useful when we come to think of what we mean by a product. If we consider the abstract statement

$$2 \times 3 = 6,$$

we cannot exactly *prove* it. We can only illustrate it by considering what it means in relation to some physical object: say a dot.

In the physical world '2 dots', or '2 lots of (a dot)' means something. Then we can take '3 lots of (2 dots)' or 3 lots of (2 lots of (a dot)), and observe that this is the same as 6 lots of (a dot).

• •
• •
• •

Now if we replace 'dots' by some other word, 'apples' or 'BMC motors', we get an equivalent result. At some stage in evolution we come, with our powers of mathematical abstraction, to suspect that replacing dots by *any* other concrete noun we get an equivalent result; and we are lead to generalize it in the formula

$$2 \times 3 = 6,$$

and on this base our definition of what we mean by multiplying two numbers.

Now if we think of it in this light, we will see that the 'product of two functions' we have defined is a case of multiplication in strictly the same sense.

If we have three functions f, g and h, and we note that

f operating on (g operating on (x))

is the same as h operating on (x)

no matter what number we put for x, then by analogy we can say that in general

$$fg = h,$$

and on this base our definition of what we mean by multiplying two functions.

But a word of caution may be needed here. When dealing with numbers it may be seen that 2×3 is the same as 3×2. This is because the figure above may be read either as '2 lots of (3 lots of (a dot))' or '3 lots of (2 lots of (a dot))', according as we reckon by columns or rows. This suggests the commutative law. There is no equivalent alternative when we interpret $f(g(x))$, and we are not justified in assuming the commutative law works here. Indeed we can prove that it does not.

For let us consider an example of such a product. If $f(x) = x^3$, and $g(x) = 1/(2x-1)$, then by definition,

$$fg(x) = f(g(x)) = f(1/(2x-1)) = 1(2x-1)^3.$$

But $gf(x) = g[f(x)] = g(x^3) = 1/(2x^3-1)$.

It is not hard to see that images of x under the two products are not the same.

Certain points may in general be made about products of functions. In the first place, the idea of product can be extended. If f, g and h are three functions, we can form a triple product, for we can consider the product of (fg) and h, or that of f and (gh), and it is easy to show that these are the same. For

$$[(fg)\, h]\, (x) = (fg)\, (h(x)) = f(g(h(x)))$$

and

$$[f(gh)]\, (x) = f(gh(x)) = f(g(h(x))).$$

We can write the product without ambiguity as fgh.

Further in the algebra of functions there is a function which plays the same part as 1 among numbers. This is the identity function, I, where $I(x) = x$; for which we can show that for all $f, fI = If = f$. For $fI(x) = f(x)$, and $If(x) = I(f(x)) = f(x)$ for all x in the domain of f.

However the main point of these products for our present purposes is that they enable us to break down apparently complicated functions into the product of simple ones whose properties we already know.

Let us take such a function, f, where $f(x) = (2x-3)^3$. If we think here of the cube function, P_3, we can write:

$$f(x) = (2x-3)^3 = P_3(2x-3)$$
$$= P_3(g(x)) \quad \text{where} \quad g(x) = 2x-3.$$

Hence
$$f = P_3 g.$$

Or take again

$$f(x) = 1/(1-2x^2)^2.$$

Here we note

$$f(x) = \frac{1}{(1-2x^2)^2} = P_{-1}[(1-2x^2)^2]$$
$$= P_{-1}P_2(1-2x^2)$$
$$= P_{-1}P_2 g(x) \quad \text{where} \quad g(x) = 1-2x^2,$$

so that
$$f = P_{-1}P_2 g.$$

If however we feel sufficiently familiar with the function P_{-2} to use it as one of our basic functions (and this will be so later on) we could analyse the function more simply as the product $P_{-2}g$.

Exercise A

1. If $f(x) = 1/x$ and $g(x) = x+1$, show using diagrams as in Figure 1, the successive mappings which lead to (i) $(fg)(-3)$; (ii) $(gf)(1)$; (iii) $(fg)(1)$; (iv) $(gf)(\frac{1}{2})$.

2. Given that $f(x) = 2x+1$ and $g(x) = x^2$, find (i) $(fg)(-1)$; (ii) $(gf)(0)$; (iii) $(fg)(x)$; (iv) $(gf)(x)$.

3. Given F such that $F(x) = \sqrt{(x^2+1)}$, find f and g such that $F(x) = f(g(x))$ where one of $f(x)$ and $g(x)$ is $\sqrt{x}$.

4. Given F such that $F(x) = (3x+7)^3$, find f and g such that $F = fg$ where one of f and g is P_3.

5. Analyse f as products of other functions which include P_2, P_3, P_{-1} and so on, where $f(x)$ equals: (i) $(3x^2-1)^2$; (ii) $1/(1-5x)$; (iii) $1/(1-x^2)^3$; (iv) $[\sqrt{(2x+1)}]^3$.

6. $f(x) = \sqrt{(x-9)}$ and $g(x) = x^2$, in each case for all real x for which the functions could be defined. What are the domains of f, g, fg and gf?

7. Given $f(x) = x-3$, $g(x) = x^2+1$, $h(x) = x+2$, find fg and gh, and hence show that $(fg)h$ and $f(gh)$ are the same function.

8. What are P_2P_3, P_3P_2, $P_{-1}P_3$ and P_3P_{-2}?

2. DERIVATIVES OF PRODUCT FUNCTIONS

2.1 Differentiating a product function. There is a simple relationship between the derived function of a product of two functions and the derived functions of its separate parts. To show this we shall first establish the graphical meaning of the derivative when the function is illustrated, as in Figure 2, on two parallel lines with the same scale on both.

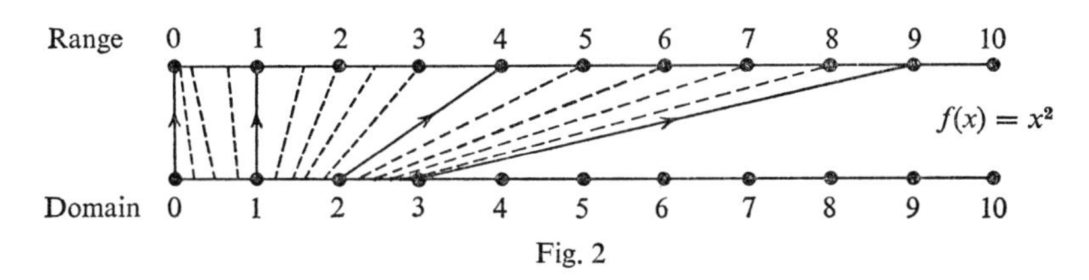

Fig. 2

As an example we shall take the function P_2. Lines connect the elements of the domain with their images. If we think of the domain-axis as a piece of elastic, on which the scale is as shown, we can imagine it being stretched so that it fits over the range-axis, with each point of the domain now lying on top of its image in the range—the 2 fitting over the 4, 3 over 9, 2·5 over 6·25, and so on. To do it with absolute accuracy is impossible, since the enlargement is not uniform within any section. Indeed, when any length of elastic is stretched to fit its end-points correctly, all intermediate points are wrongly placed. For instance, the interval from 2 to 3 must be enlarged by a factor 5; but the interval from 2 to 2·1 is enlarged by a factor 4·1 (since $2^2 = 4$ and $2{\cdot}1^2 = 4{\cdot}41$, and therefore a length 0·1 is enlarged to a length 0·41). Furthermore, the interval from 2 to 2·01 is enlarged by a factor 4·01, and so on. This process of considering the enlargements for shorter and shorter intervals starting from a particular point leads us to a definition of the *enlargement at the point*.

Consider a general function f. x maps onto $f(x)$ and $x+\epsilon$ onto $f(x+\epsilon)$, so that an interval of length ϵ maps onto one of length

$$f(x+\epsilon)-f(x).$$

This gives us an enlargement for the interval of

$$[f(x+\epsilon)-f(x)]/\epsilon.$$

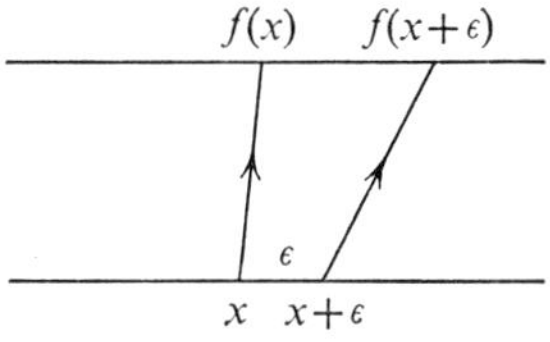

$f(x)$ $f(x+\epsilon)$

ϵ

x $x+\epsilon$

Fig. 3

Now if f is a differentiable function,

$$f(x+\epsilon) = f(x)+f'(x).\epsilon+s_\epsilon.$$

Hence the enlargement for the interval is $f'(x)+s_\epsilon/\epsilon$. But this can be made as close to $f'(x)$ as we like, by taking short enough interval of length ϵ. We therefore define the enlargement of the mapping *at* the point to be $f'(x)$.

We can now apply this idea to a composite function, and as an example we shall find $F'(1)$, where $F(x) = (3x-1)^2$.

Now $F = fg$, where $f(x) = x^2$ and $g(x) = 3x-1$. g maps 1 onto 2, and then f maps 2 onto 4. The enlargement for g at 1 is $g'(1)$ which is 3. That for f at 2 is $f'(2)$, which is 4. We can think therefore of the line A being enlarged by a factor 3 at X to fit it over B at Y, and the line B being enlarged by a factor 4 at Y to fit it over C at Z.

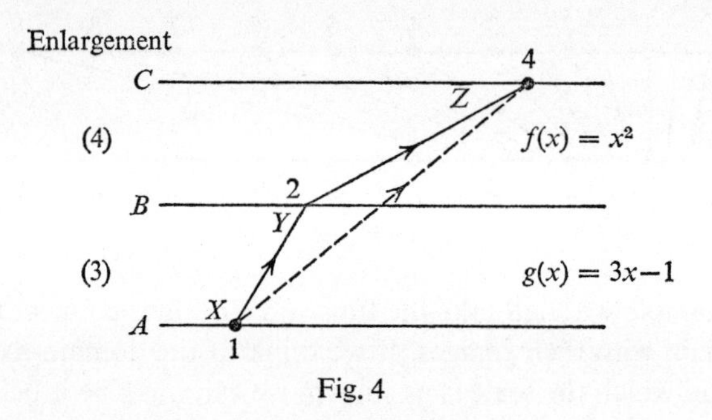

Fig. 4

Hence the enlargement of A at X necessary to fit it over C at Z is the produce of 3 and 4, and it follows that $F'(1) = 12$.

(That this is indeed so can be seen by writing $F(x) = 9x^2-6x+1$, so that $F'(x) = 18x-6$ and $F'(1) = 18-6 = 12$.)

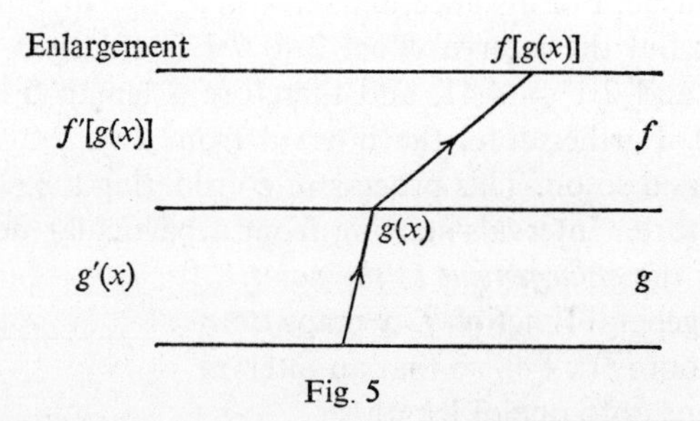

Fig. 5

The general argument is illustrated in Figure 5. Suppose that $F = fg$, where f and g are differentiable functions. We argue that g maps x onto $g(x)$ and then f maps $g(x)$ onto $f[g(x)]$. The enlargements are $g'(x)$ at x and $f'[g(x)]$ at $g(x)$, so that we expect

$$F'(x) = f'[g(x)]\,g'(x).$$

The reader may possibly think that the argument used is conclusive; but although it is obvious that if a *finite interval* were enlarged first by a factor 3 and then by a factor 4, all in all it would be enlarged by a factor 12, we cannot assume that enlargements *at points* behave so conveniently. For such an enlargement is a sophisticated idea. As there is a matter of technical difficulty in a rigorous proof we shall content ourself with a statement of the theorem only.

If f and g are differentiable functions, then fg is a differentiable function, with derived function defined by the equation:

$$(fg)'(x) = f'[g(x)]\, g'(x).$$

Example 1. Find the derivative of $F(x) = (x^2+1)^3$.

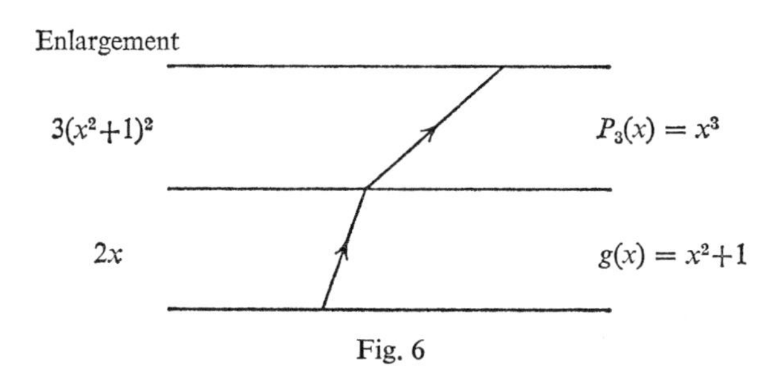

Fig. 6

Here $\quad F = P_3 g \quad$ where $\quad g(x) = x^2+1.$

Now $\quad P_3' = 3P_2.$

Therefore $\quad P_3'(x^2+1) = 3(x^2+1)^2.$

Also $\quad g'(x) = 2x.$

Hence $\quad F'(x) = 3(x^2+1)^2\, 2x = 6x(x^2+1)^2.$

Example 2. Find the derivative of $F(x) = 1/(x^2-3x)$.

F is the product of P_{-1} and the function under which x maps onto x^2-3x.

The corresponding enlargements are $-(x^2-3x)^{-2}$ and $2x-3$. Hence

$$F'(x) = -(x^2-3x)^{-2}\,(2x-3).$$

Example 3. Find the derivative of $F(x) = [g(x)]^n$ in terms of $g(x)$ and $g'(x)$.

Here $F = P_n g$. The corresponding enlargements are $n[g(x)]^{n-1}$ and $g'(x)$. Hence

$$F'(x) = n[g(x)]^{n-1}\, g'(x).$$

Once the reader understands this result, he should be able to write down the derivative of such expressions without recourse to a diagram or naming the functions which are combined. The following examples are intended to help him check his ability to write down the derivatives directly.

$F(x)$	$F'(x)$
$(2x+1)^5 = P_5(2x+1)$	$5(2x+1)^4\,2 = 10(2x+1)^4$
$\dfrac{1}{x^3} = (x^3)^{-1} = P_{-1}(x^3)$	$-(x^3)^{-2}\,3x^2 = -\dfrac{3}{x^4}$
$\dfrac{1}{x^3} = (x^{-1})^3 = P_3(x^{-1})$	$3(x^{-1})^2\,(-x^{-2}) = -\dfrac{3}{x^4}$
$(1-x)^4 = P_4(1-x)$	$4(1-x)^3\,(-1) = -4(1-x)^3$
$(x^3+3x^2-2)^3 = P_3(x^3+3x^2-2)$	$3(x^3+3x^2-2)^2\,(3x^2+6x)$ $= 9x(x+2)\,(x^3+3x^2-2)^2.$

Exercise B

1. For the function $f(x) = x^3+1$, find the enlargements under the mapping of the intervals from 1 to (i) 2; (ii) 1·5; (iii) 1·1; (iv) 1·01; (v) 1·001; and comment on your results.

2. If the function ϕ is the product of the functions P_2 and g, where $g(x) = x^3-5$, show in a figure, like Figure 4, the successive mappings of 2 under g followed by P_2. Calculate $P_2'(3)$ and $g'(2)$, and hence find $\phi'(2)$. Write down $P_2'(x^3-5)$ and $g'(x)$, and hence find $\phi'(x)$.

3. In each of the following draw a figure to show the product mappings and the two enlargements, as in Figure 4, and hence find:
(i) $\phi'(2)$ where $\phi(x) = (\frac{1}{2}x+1)^3$; (ii) $\phi'(3)$ where $\phi(x) = (4-x)^2$, explaining the significance of the sign in relation to the mapping.

4. In each of the following draw a figure to show the product mappings and the two enlargements as in Figure 6, and hence differentiate:

(i) $(3x+1)^2$; (ii) $(x^2+3x-1)^5$; (iii) $(5-2x)^4$;
(iv) $1/(x^2+7)^4$; (v) $1/(1-x)^2$.

5. Use the fact that $P_6 = P_2P_3$, to find the derivative of x^6.

6. Express the function P_{-4} as a product of functions, and hence find its derivative. Repeat using a different composition.

7. Write down the derivatives of the following and then check by multiplying out: (i) $(1-3x)^2$; (ii) $(x^2+1)^3$; (iii) $(2x-x^3)^2$.

8. Write down the derivatives of:

(i) $(2x-3)^4$; (ii) $(3x^2-3x+2)^5$; (iii) $(1-2x)^3$;
(iv) $(x^2+1)^{-1}$; (v) $1/x^5$; (vi) $(1+x^4)/x^3$.

9. Sketch the graph of $f(x) = \dfrac{1}{x}+\dfrac{1}{x^2}$ and find its turning points.

10. Sketch the graph of $f(x) = \dfrac{1}{4x^2}+\dfrac{1}{(2x-1)^2}$ and find its turning points.

11. A right circular cone has a volume of $\pi/3\ \text{m}^3$. What is its least possible curved surface area? [C.S.A. $= \pi rl$, where l is the slant height; express the square of the C.S.A. as a function of the height.]

12. A vase is of external radius $81/(x^2-6x+36)$ cm at height x cm, where $0 \leqslant x \leqslant 10$. Sketch its shape. Find where it is widest and the angle of slope near its base.

13. Establish a rule for the derivative of the triple product fgh of functions f, g, h. Use the rule to find the derivative of $\left[\dfrac{1}{(x^2+3)}\right]^3$, stating carefully what functions are here represented by f, g, h.

14. Show that the enlargement of the interval from x to $x+\epsilon$ under the function P_{-1} is $-\dfrac{1}{x(x+\epsilon)}$. Deduce the derivative of x^{-1}.

3. DERIVATIVES OF RATIONAL POWERS OF x

We have already seen how to differentiate x^{-2}, x^{-3} and so on. A general rule for differentiating negative integral powers of x will now be established which can be extended to any rational power of x.

3.1 Derivative of x^{-m} (where m is a positive integer). If $f(x) = x^{-m}$, we can write $f(x) = (x^{-1})^m$, so that

$$f'(x) = m(x^{-1})^{m-1}(-x^{-2}) = -mx^{-m-1}.$$

We can write this: $$P'_{-m} = -mP_{-m-1}.$$

The rule $P'_n = nP_{n-1}$, which has been shown to hold when n is a positive integer, zero, or -1, is now seen to be valid also for all negative integers.

3.2 Derivative of $x^{p/q}$ (p and q integers). If $f(x) = x^{p/q}$, then $\{f(x)\}^q = x^p$.

Equating their derivatives,

$$q\{f(x)\}^{q-1}f'(x) = px^{p-1};$$

and multiplying both sides by $f(x)$, we have

$$q\{f(x)\}^q f'(x) = px^{p-1}f(x).$$

That is $$qx^pf'(x) = px^{p-1}x^{p/q},$$

and therefore $$f'(x) = \frac{p}{q}\,x^{(p/q)-1}.$$

This result shows that the rule $P'_n = nP_{n-1}$, holds for all rational values of n.

For example, the derivative of $x^{\frac{2}{3}}$ is $\frac{2}{3}x^{-\frac{1}{3}}$; the derivative of $x^{-\frac{5}{2}}$ is $-\frac{5}{2}x^{-\frac{7}{2}}$; and the derivative of $\sqrt{x}$, that is $x^{\frac{1}{2}}$, is $\frac{1}{2}x^{-\frac{1}{2}}$ or $1/2\sqrt{x}$.

Example 1. Differentiate $1/\sqrt{(2x-3)^3}$.

This can be written $(2x-3)^{-\frac{3}{2}}$, and using the rule for the derivative of the product of functions, the derivative is

$$-\tfrac{3}{2}(2x-3)^{-\frac{5}{2}}\,2 = -3/\sqrt{(2x-3)^5}.$$

Exercise C

1. Differentiate: (i) $x^{\frac{1}{2}}$; (ii) $x^{-\frac{1}{2}}$; (iii) $\sqrt[3]{x}$; (iv) $\sqrt{(3x-7)}$; (v) $1/\sqrt{(3-x)}$.

2. Differentiate: (i) $3x^{\frac{2}{3}}$; (ii) $7\sqrt[3]{x}$; (iii) $5\sqrt{(4x^2-3)}$; (iv) $6/\sqrt{(2-3x)}$.

3. Sketch the graph of $f(x) = \sqrt{x^3}$. If f is defined for all possible values of x, what is its domain? What is $f'(0)$?

4. Sketch the graph of $f(x) = (x+1)/\sqrt{x}$ by adding the graphs of $\sqrt{x}$ and $1/\sqrt{x}$. Find its turning point. Find also $f'(4)$ and $f'(9)$; which is the greater?

5. Sketch the graph of $f(x) = 1/\sqrt{(1-x)}$. With the same axes, and using a broken line, sketch the graph of $f'(x)$. Find $f'(x)$.

6. If $f(x) = \sqrt{x}$, find, by using $f(x+\epsilon) \approx f(x)+f'(x)\epsilon$, an approximation to $\sqrt{(9+\epsilon)}$. Check, by squaring, that your approximation is reasonable.

7. If $f(x) = x^{-1}$, find, by using $f(x+\epsilon) \approx f(x)+f'(x)\epsilon$, an approximation to $1/(10+\epsilon)$.

8. Use the method of linear approximation to get approximate values for (i) $\sqrt{(9{\cdot}3)}$; (ii) $1/\sqrt{(24{\cdot}8)}$.

9. Sketch the graph of (i) $f(x) = x-\sqrt{(1+2x)}$; (ii) $g(x) = x-\sqrt{(1+x^2)}$.

10. Two roads meet at right angles at O. A man A walks along one road at 4/3 m/s and at the instant when A is at O, a man B is 200 m away on the other road approaching at 1 m/s. Find an expression for their distance apart after t sec. For how many seconds does the distance between them decrease?

4. DERIVATIVE OF $f(x)\,g(x)$

For functions of the form $f(x)\,g(x)$ (such as $x\sqrt{(2x+1)}$) it is sometimes inconvenient or impossible to put them into a form which we are now able to differentiate. The following theorem allows us to deal with such functions without modification.

4.1 Derivative of $f(x)\,g(x)$. If f and g are differentiable functions then the derivative of $f(x)\,g(x)$ is

$$f'(x)\,g(x)+f(x)\,g'(x).$$

Suppose F is a function such that $F(x) = f(x)\,g(x)$. Then

$$F(x+\epsilon) = f(x+\epsilon)\,g(x+\epsilon).$$

Now f and g are differentiable.

Hence $$f(x+\epsilon) = f(x)+f'(x)\,\epsilon+s_\epsilon$$

and $$g(x+\epsilon) = g(x)+g'(x)\,\epsilon+s_\epsilon.$$

Therefore

$$F(x+\epsilon) = [f(x)+f'(x)\epsilon+s_\epsilon]\,[g(x)+g'(x)\,\epsilon+s_\epsilon]$$

$$= F(x)+[f'(x)\,g(x)+f(x)\,g'(x)]\,\epsilon+s_\epsilon.$$

Hence $$F'(x) = f'(x)\,g(x)+f(x)\,g'(x).$$

This proof depends on the fact that terms small compared with ϵ multiplied by other terms small compared with ϵ give terms small compared with ϵ. This is not hard to prove from our original definition of such terms in Chapter 3. Indeed the reader may be content to accept it as obvious.

Example 1. Find the derivative of $F(x) = x\sqrt{(3x+1)}$.

Here

$$F(x) = f(x)\,g(x) \quad \text{where} \quad f(x) = x \quad \text{and} \quad g(x) = \sqrt{(3x+1)}.$$

With $f'(x) = 1$ and $g'(x) = \frac{1}{2}(3x+1)^{-\frac{1}{2}}\,3$, we can use the formula to give

$$F'(x) = 1(3x+1)^{\frac{1}{2}}+x\tfrac{3}{2}(3x+1)^{-\frac{1}{2}}.$$

This expression can be put in simpler form by extracting the factor $\frac{1}{2}(3x+1)^{-\frac{1}{2}}$ from each term, where $(3x+1)^{-\frac{1}{2}}$ is the *lowest* power of $(3x+1)$ in the two terms.

This gives us something of the form $\frac{1}{2}(3x+1)^{-\frac{1}{2}}\,[A+B]$. We now ask ourselves what we can write for A and B so that when multiplied by $\frac{1}{2}(3x+1)^{-\frac{1}{2}}$ they give $(3x+1)^{\frac{1}{2}}$ and $x\frac{3}{2}(3x+1)^{-\frac{1}{2}}$ respectively.

We get

$$F'(x) = \tfrac{1}{2}(3x+1)^{-\frac{1}{2}}\,[2(3x+1)+3x]$$

$$= \tfrac{1}{2}(3x+1)^{-\frac{1}{2}}\,(9x+2).$$

Alternatively, some readers may prefer to simplify $F'(x)$ by writing it in fractional form:

$$F'(x) = \sqrt{(3x+1)}+\frac{3x}{2\sqrt{(3x+1)}} = \frac{2(3x+1)+3x}{\sqrt{(3x+1)}} = \frac{9x+2}{2\sqrt{(3x+1)}}.$$

It is often the tidying up process which is most difficult.

Example 2. Find the derivative of $F(x) = (x-1)^3 (3x^2-1)^4$.

$$\begin{aligned} F'(x) &= 3(x-1)^2 (3x^2-1)^4 + (x-1)^3 4(3x^2-1)^3 6x \\ &= (x-1)^2 (3x^2-1)^3 [3(3x^2-1)+4(x-1)(6x)] \\ &= (x-1)^2 (3x^2-1)^3 (9x^2-3+24x^2-24x) \\ &= (x-1)^2 (3x^2-1)^3 (33x^2-24x-3) \\ &= 3(x-1)^2 (3x^2-1)^3 (11x^2-8x-1). \end{aligned}$$

Example 3. Sketch the graph of $x\sqrt{(1-2x)}$.

Let
$$f(x) = x\sqrt{(1-2x)}.$$

We notice first that $f(x)$ only exists when $x < \frac{1}{2}$, otherwise $1-2x$ is negative.

Now when x is small $f(x) \approx x$, so that the graph passes through the origin with unit gradient. When x is large and negative $f(x) \approx x\sqrt{(-2x)}$ which gives the shape of graph shown in the branch which curves downwards to infinity.

When $x = \frac{1}{2}$, $f(x) = 0$; and to find the shape of the graph there we need to know $f'(x)$.

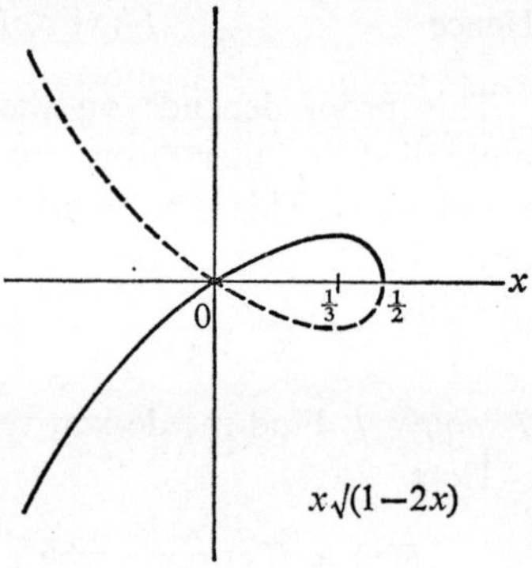

Fig. 7

Now $f(x) = x(1-2x)^{\frac{1}{2}}$.

So that
$$\begin{aligned} f'(x) &= x\tfrac{1}{2}(1-2x)^{-\frac{1}{2}}(-2)+(1-2x)^{\frac{1}{2}} \\ &= -x(1-2x)^{-\frac{1}{2}}+(1-2x)^{\frac{1}{2}} \\ &= (1-3x)/(1-2x)^{\frac{1}{2}}, \end{aligned}$$

and when x tends to $\frac{1}{2}$, $f(x)$ tends to $-\infty$, so that the graph becomes vertical at $x = \frac{1}{2}$. We can also check that $f'(0) = 1$ to confirm the gradient at the origin. Furthermore, we observe that $f(x)$ has a greatest value where $x = \frac{1}{3}$.

Exercise D

1. Find the derivatives of the following: (*a*) using the method of Section 4.1; (*b*) by first multiplying out: (i) $(2x+3)(x^2+1)$; (ii) $x^2(1-x^3)$; (iii) $\frac{1}{x}(x^2+1)$; (iv) $x.x^6$.

2. Find the derivatives of the following *without* multiplying out first: (i) $x^2(2x-1)^3$; (ii) $(1-x)^2(1+x)^3$; (iii) $x(4x^2+2x-1)^3$.

3. Differentiate:

(i) $x^2.\sqrt{(x+1)}$; (ii) $x/\sqrt{(x+1)}$; (iii) $\sqrt{(x^2-3x-1)}$;
(iv) $\sqrt{\{(x+1)(1-2x)\}}$; (v) $(x^2-3)/\sqrt{(x+1)}$.

4. Sketch the graph of $f(x) = x^2\sqrt{(1-x)}$, and find the turning values. If f is defined for all possible values of x, what is its domain? Show in the same figure, using a broken line, the graph of $g(x) = -x^2\sqrt{(1-x)}$.

5. Sketch the graph of $f(x) = x\sqrt{(x^2-1)}$. Find the points of inflection.

6. Sketch the graph of $x/(x^3-2)$. How many real roots has $x^3-2x-2 = 0$?

7. An isosceles triangle is inscribed in a circle of radius a cm. Show that it has the maximum area when it is equilateral. [Let x cm be the perpendicular distance from the centre to the base; express the area in terms of x and a.]

8. Write down the distance from the origin to the point (λ, μ) in the Cartesian plane. If (λ, μ) lies on the line $3x-2y+1 = 0$, express the distance in terms of either λ or μ alone. Hence show that the least distance occurs along the perpendicular to the line.

9. Find the height of the circular cylinder of maximum curved surface area which can be inscribed in a sphere of unit radius.

10. An aircraft should be flying on a track 060°, helped by a south wind of speed w units. If the airspeed of the aircraft is 1 unit, show that it will travel, with respect to the ground, at $(1-\frac{3}{4}w^2)^{\frac{1}{2}}+\frac{1}{2}w$ units, provided it flies on this track at all. Hence find (i) for what values of w it is possible to fly on this track, and (ii) for what value of w it flies on it with greatest speed.

11. A length of wool runs from a ball on a table over the edge (crossing it at right angles) so that the free end is just touching the floor. A cat grabs this end and backs directly away from the table at 1 m/s. If the height of the table is 1 m, find in terms of the time t seconds from when the cat starts, the length of wool from the table edge to the cat. At what speed is the wool crossing the edge after 2 seconds?

12. A bomb explodes at a height h m above level ground and the shock wave moves outwards at V m/s. At what rate, in terms of V, h, t, does the radius of the circular intersection of the shock wave with the ground increase at time t seconds *after* the explosion? If this rate is denoted by u m/s, so that this is the speed of advance of the circular shock wave front over the ground, what is the front's outward acceleration? Sketch the general shape of the graph of u plotted against t, starting at the instant when the shock wave meets the ground.

13. A police launch moves on a track which makes an angle of $x°$ with the direction of the current of a river. The speed of the current is 1 unit, and that of the launch relative to the water is 2 units. Express the speed of the launch as a function of $\cos x°$, and hence sketch the graph of this speed against x in the range 0° to 360°.

14. Two long rods AB and BC are jointed at B and pass through rings E and F in fixed positions at the same level 2 m apart, one through each ring. B moves up and down the vertical line through the mid-point of EF. Express the height above EF of the point P on BC, where BP is d m, in terms of $\cos x°$, where BC makes an angle $x°$ with the downward vertical. Hence sketch the graphs of the height as a function of x for the cases $d = \frac{1}{2}$, 1 and 2, over the range 0° to 180°.

15. A calculating but lazy shark sights an approaching Channel swimmer 5 cables away when he is 3 cables from the swimmer's track. He estimates the speed of the swimmer to be 2 knots. He feels indisposed to swim at more than 1 knot. Is it worth his while to give chase to the swimmer? (Express the shark's necessary interception speed as a function of x, where the point of interception is $4+x$ cables beyond the original position of the swimmer.)

16. $ABCD$ is a rectangular ploughed field with a path along each edge. AB is 400 m and AD 100 m long. A boy, who can walk at 2 m/s along the path but at only 4/3 m/s across the field, wants to walk from A to the mid-point of CD as quickly as possible. What route should he take?

5. MISCELLANEOUS

5.1 Quotients. To find the derivative of a quotient, for instance of $F(x) = x/(x^2+1)$, we can write it in the form $x(x^2+1)^{-1}$, and use the rule for differentiating a product of two functions of x.

Thus
$$F(x) = x(x^2+1)^{-1},$$
so that
$$\begin{aligned} F'(x) &= (x^2+1)^{-1} - x(x^2+1)^{-2}\, 2x \\ &= (x^2+1)^{-2}\,(x^2+1-2x^2) \\ &= (x^2+1)^{-2}\,(1-x^2). \end{aligned}$$

We can also however prove a formula for the derivative of a quotient, which some readers may prefer to use.

Let
$$F(x) = f(x)/g(x) = f(x)\,[g(x)]^{-1}.$$
Then
$$\begin{aligned} F'(x) &= f'(x)\,[g(x)]^{-1} - f(x)\,[g(x)]^{-2}\,g'(x) \\ &= [g(x)]^{-2}\,[f'(x)\,g(x) - f(x)\,g'(x)] \end{aligned}$$
or,
$$F'(x) = \frac{f'(x)\,g(x) - f(x)\,g'(x)}{[g(x)]^2}.$$

In our previous example, $F(x) = x/(x^2+1)$. Hence
$$F'(x) = \frac{(x^2+1)\,1 - x.2x}{(x^2+1)^2} = (1-x)^2/(x^2+1)^2.$$

5.2 A note on function notation. We have already a function notation for the product of two functions which we discussed in Section 1.1. There, if we were multiplying functions f and g we wrote the product fg.

We can introduce similar notations for those kinds of product function we have been discussing in Section 4.1. Let $f.g$ denote the function which maps x onto $f(x)\,g(x)$. Thus $f.g$ is defined by the equation
$$(f.g)\,(x) = f(x)\,g(x).$$

Likewise we can introduce functions $f+g$, $f-g$, and f/g defined by the equations

$$(f+g)(x) = f(x)+g(x),$$

$$(f-g)(x) = f(x)-g(x),$$

$$(f/g)(x) = f(x)/g(x).$$

Using these definitions (and being careful not to confuse fg and $f.g$) we can write our rules for differentiating composite functions:

$$(fg)' = f'g.g',$$

$$(f.g)' = f'.g+f.g',$$

$$(f/g)' = (f'.g-f.g')/(g.g).$$

An obvious extension of the notation is to write $f+f = 2f$, and in general define kf (where k is a constant) by the equation $(kf)(x) = kf(x)$.

We could write then another (though simpler) rule, corresponding to those we developed in Chapter 3:

$$(pf+qg)' = pf'+qg'.$$

The reader will be able to see that we might, with those rules, develop a whole algebra of functions, without ever mentioning individual elements of their domains, or variables. To do so is beyond the range of this book.

Exercise E

1. Differentiate: (i) $(1+x)/(1+x^2)$; (ii) $2x/(x^2-1)$; (iii) $(x-1)^2/(1-2x)^2$.

2. Differentiate: (i) $x^2/\sqrt{(1+x^2)}$; (ii) $\sqrt{[(2x-1)/(x^2-1)]}$; (iii) $\sqrt{(1-4x)}/(1+2x)$.

3. Write down the derived function of $F.h$. If $F = f.g$, establish the rule

$$(f.g.h)' = f'.g.h+f.g'.h+f.g.h'.$$

Apply this rule to find the derivative of x^3 considered as $x.x.x$.

4*. Is the expression $fg.h$ ambiguous (where f, g and h are functions)? Using only the notation of Section 5.2 write down the derived functions of as many different interpretations as you can give it.

7

EXPONENTIAL FUNCTION

So far in this book we have dealt only with algebraic functions: functions, that is, which can be built up from the simple powers like x^3 and $x^{-\frac{1}{2}}$. In power functions we find always a variable raised to the power of a constant; but these are not the only type. We have only to think for instance of f where $f(x) = 3^x$, in which we have a constant raised to the power of a variable, to get something radically different. Functions of this kind we call *exponential functions.* They have a fundamental importance in mathematics.

1. a^x

1.1 $f(x) = 2^x$. It will be helpful to start with a project.

Make out a table of values for 2^x for $x = -2, -1, 0, 1, 2, 3, 4$.

Note from this that wherever we increase x by 1, 2^x is doubled.

Prove that this is a general rule. (Use the law of indices.)

Note that $2^{\frac{1}{2}} = \sqrt{2} = 1{\cdot}414$ (to three decimal places). Use this and the rule above to find 2^x for $x = 1{\cdot}5, 2{\cdot}5, 3{\cdot}5$ and for $x = -0{\cdot}5, -1{\cdot}5$. (For instance, $2^{1\cdot 5} = 2 \times 2^{0\cdot 5} = 2 \times 1{\cdot}414 = 2{\cdot}828$.) Hence enlarge your table of values, giving them correct to two decimal places.

Taking the scales of 1 inch = 1 unit for the domain, and 1 inch = 2 units for the range, draw as accurately as you can the graph of 2^x from $x = -2$ to $x = 4$.

Draw tangents to this graph at $x = 0, 1, 2, 3$; and hence find values for $f'(0), f'(1), f'(2)$ and $f'(3)$ correct to one decimal place.

Tabulate these values together with those of $f(x)$.

Can you see any pattern in your table? If so, check it by finding $f'(-1)$.

Let us summarize the results that you should have got. The table of values should be:

x	-2	$-1{\cdot}5$	-1	$-0{\cdot}5$	0	$0{\cdot}5$	1
2^x	0·25	0·35	0·50	0·71	1·00	1·41	2·00

x	1·5	2	2·5	3	3·5	4
2^x	2·83	4·00	5·66	8·00	11·31	16·00

The graph, reduced in scale, appears in Figure 1. We see that $2^x \to \infty$ as $x \to \infty$.

The estimated values of $f'(x)$ should have been

x:	0	1	2	3
$f'(x)$:	0·7	1·4	2·8	5·5

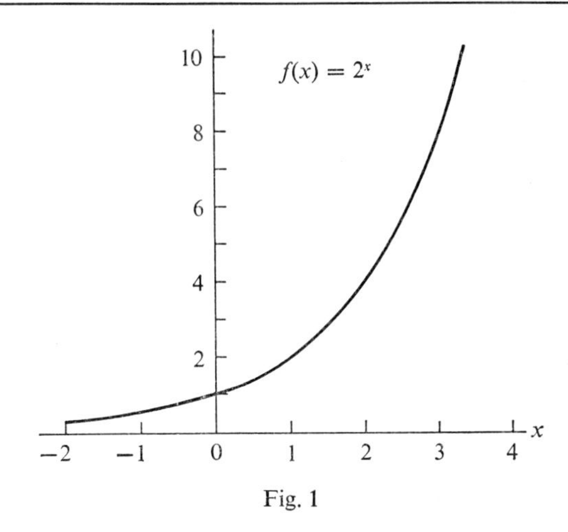

Fig. 1

Clearly there is a pattern in these. Each number seems to be twice the last: except 5·5 that is, but the discrepancy may be unreal. For these were found correct to one decimal places. Suppose that more accurate values were found to be:

$$0{\cdot}688, \quad 1{\cdot}376, \quad 2{\cdot}752, \quad 5{\cdot}504,$$

where each is exactly twice the last. There values corrected to one decimal place would have been those given above.

Let us then take it as a working hypothesis that the values of $f'(x)$ show this pattern and tabulate them in the form:

x:	0	1	2	3
$f'(x)$:	$0{\cdot}7\times 2^0$	$0{\cdot}7\times 2^1$	$0{\cdot}7\times 2^2$	$0{\cdot}7\times 2^3$.

This at least suggests that for such values of x the gradients are proportional to the heights of the graph: in other words, that the rate of change of 2^x is proportional to 2^x itself.

We can repeat this work for some other exponential function. For instance, with $f(x) = 3^x$, the reader can verify that

x:	0	1	2	3
$f(x)$:	3^0	3^1	3^2	3^3
$f'(x)$:	$1{\cdot}1\times 3^0$	$1{\cdot}1\times 3^1$	$1{\cdot}1\times 3^2$	$1{\cdot}1\times 3^3$.

Here again it seems that the rate of change of 3^x is proportional to 3^x, although now the constant of proportionality is 1·1, whereas before it was only 0·7. Obviously then, even allowing for the inevitable inaccuracies of results estimated by eye, there is a case for suspecting that the derivative of a^x is proportional to a^x, and that the constant of proportionality depends upon the constant a. We shall now prove this to be true. (Our proof depends upon the assumption that the function a^x has got a proper derivative at the particular point $x = 0$, but this the reader will have no difficulty in accepting from the smoothness of the graph.) We will prove it for $a = 2$, but it will be clear that the general result could be established in a similar way.

1.2 Derivative of 2^x. Let $f(x) = 2^x$, and, assuming that f is differentiable where $x = 0$, let $f'(0) = k$.

We wish to express $f(x+\epsilon)$ in the form

$$f(x+\epsilon) = f(x)+\epsilon f'(x)+s_\epsilon. \qquad \text{(i)}$$

Now $$f(\epsilon) = f(0)+\epsilon f'(0)+s_\epsilon.$$

But $$f(\epsilon) = 2^\epsilon;\quad f(0) = 2^0 = 1 \quad\text{and}\quad f'(0) = k;$$

hence $$f(\epsilon) = 2^\epsilon = 1+k\epsilon+s_\epsilon.$$

Now $$f(x+\epsilon) = 2^{x+\epsilon} = 2^x 2^\epsilon,$$

So that $$\begin{aligned} f(x+\epsilon) &= 2^x(1+k\epsilon+s_\epsilon) \\ &= 2^x+k2^x\epsilon+2^x.s_\epsilon \\ &= f(x)+k2^x\epsilon+s_\epsilon. \end{aligned}$$

Hence, comparing this with (i), we see that

$$f'(x) = k2^x.$$

Here then is our result. We see that the derivative of 2^x is proportional to 2^x, and the constant of proportionality is its derivative where $x = 0$, or the gradient of its graph where $x = 0$.

1.3 Derivative of a^x. Now clearly we can use this method to differentiate any function f where $f(x) = a^x$ and a is a positive constant. We will get the result:

$$f'(x) = Ka^x,$$

where K is the gradient of its graph where $x = 0$. Evidently this is a constant which depends only on a. Let us write it k_a, so that in general if $f(x) = a^x, f'(x) = k_a a^x$, or $f'(x) = k_a f(x)$.

We saw by measurement that $k_2 \approx 0{\cdot}7$; and we cannot at the moment do better than measuring to determine these constants. We can, however, establish various relations between them.

Consider, for instance, the obviously related group of functions of x, 2^x, 4^x, 8^x; and their reciprocals $(\frac{1}{2})^x$, $(\frac{1}{4})^x$, $(\frac{1}{8})^x$. Their graphs are shown together in Figure 2.

0 always maps onto 1; but if $x > 0$, $8^x > 4^x > 2^x$; and if $x < 0$, $8^x < 4^x < 2^x$. (For instance $8^{-2} = \frac{1}{64}$, $4^{-2} = \frac{1}{16}$, $2^{-2} = \frac{1}{4}$.)

In addition the graphs of (for example) 4^x and $\frac{1}{4}^x$ are symmetrical about the range-axis, as we see from the table of values:

x:	-2	-1	0	1	2
4^x:	$\frac{1}{16}$	$\frac{1}{4}$	1	4	16
$\frac{1}{4}^x$:	16	4	1	$\frac{1}{4}$	$\frac{1}{16}$

(It also follows from the fact that $\frac{1}{4}^x = 4^{-x}$.)

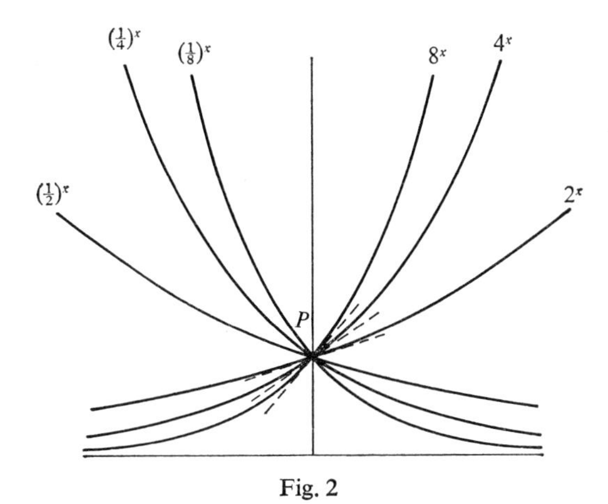

Fig. 2

All this gives us the sketch in Figure 2; and clearly, since k_a is the gradient of the graph of a^x where $x = 0$, $k_2 < k_4 < k_8$, and $k_{\frac{1}{2}} = -k_2$, $k_{\frac{1}{4}} = -k_4$, and so on.

Thus if $f(x) = \frac{1}{2}^x$, $f'(x) = -0{\cdot}7 \times \frac{1}{2}^x$.

This need not surprise us, since the derivative of such a function is clearly, from its graph, always negative.

Let us go a little deeper into this relationship. Consider

$$F(x) = 8^x.$$

Now $8 = 2^3$, so that we can write

$$F(x) = (2^3)^x = 2^{3x}.$$

Regarding F as the product of the two functions f and g where $f(x) = 2^x$, $g(x) = 3x$, we get:

$$F'(x) = f'g(x)\,g'(x) = k_2 2^{3x} 3 = 3k_2 8^x.$$

Hence

$$k_8 = 3k_2.$$

This result can be generalized. It is left to the reader to prove that $k_4 = 2k_2$, $k_{\frac{1}{2}} = -k_2$, $k_{\frac{1}{4}} = -2k_2$ and so on.

And indeed, for any positive a, k_a can be expressed in terms of k_2 by noting that

$$a = 2^{\log_2 a} \quad \text{(from the definition of logarithms)}$$

so that

$$k_a = (\log_2 a)\, k_2.$$

Likewise it could have been expressed in terms of k_3—even if there is no particular advantage in using 2 or 3 as the basic number with which to work in this way. The basic number we use is discussed in a later section.

Another way in which the constants are related can be suggested by considering the constant for 6 and its relation to those for 2 and 3; for

$$f(x) = 6^x = 2^x 3^x$$

implies

$$\begin{aligned} f'(x) &= k_2 2^x 3^x + 2^x k_3 3^x \\ &= (k_2 + k_3)\, 2^x 3^x \\ &= (k_2 + k_3)\, 6^x, \end{aligned}$$

as we see by using the formula for differentiating a product. Hence the constant for a product of two numbers equals the sum of their individual constants: a result from which we could derive some of those that are proved above.

So we can build up a repertory of constants in the work that follows by using a few measured results. The measured values of k_2, k_3 and k_5 are, to one decimal place, 0·7, 1·1 and 1·6.

1.4 Graph sketching.

Example 1. Sketch the graph of $f(x) = 3^x - 2^x$.

Sketching the graphs of 2^x and 3^x, we see that if $x > 0$, 3^x quickly becomes much the larger. Hence as $x \to \infty$, $f(x) \to \infty$.

If on the other hand $x < 0$, $3^x < 2^x < 1$. Hence $f(x)$ is negative, though it clearly tends to 0 as $x \to -\infty$.

$$f'(x) = 1{\cdot}1 \times 3^x - 0{\cdot}7 \times 2^x.$$

Hence

$$f'(x) = 0 \quad \text{where} \quad 2^x/3^x = 1{\cdot}1/0{\cdot}7.$$

Taking logarithms of both sides, and noting that $2^x/3^x = (\frac{2}{3})^x$, we get

$$x \log (\tfrac{2}{3}) = \log 1{\cdot}1 - \log 0{\cdot}7.$$

From this we find there is a minimum where $x \approx -1{\cdot}2$.

(All the numerical working in these examples is, of course, approximate, since we are using approximate values for k_a.)

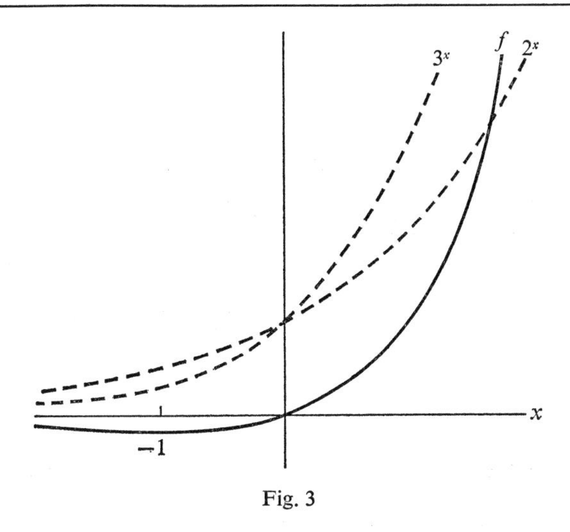

Fig. 3

Example 2. Sketch the graph of $x(\frac{1}{2})^x$.

Here we must multiply the two graphs of x and $(\frac{1}{2})^x$. The most obvious puzzle is what happens as $x \to \infty$. For one part of the product (x) tends to ∞, while the other part ($(\frac{1}{2})^x$) tends to zero.

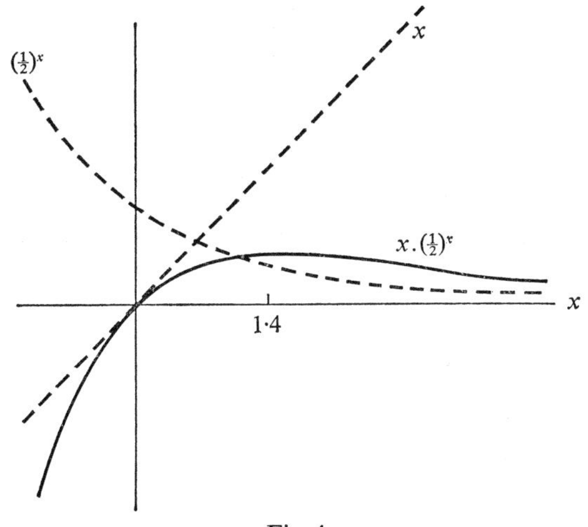

Fig. 4

A similar problem would arise if we multiplied x by x^{-2} (where the product tends to zero); or if we multiplied x^2 by x^{-1} (where it tends to ∞). Clearly then there is no standard answer to such problems, each of which must be investigated in its own right.

Here
$$f'(x) = (\tfrac{1}{2})^x - 0{\cdot}7x(\tfrac{1}{2})^x = (1-0{\cdot}7x)\,(\tfrac{1}{2})^x.$$

Hence $f'(x) \lesseqgtr 0$ according as $x \gtreqless 1/0{\cdot}7 = 1{\cdot}4$ (to one decimal place). This shows that we have a maximum where $x \approx 1{\cdot}4$. Beyond this maximum $f(x)$ is clearly decreasing, though it never becomes negative.

It seems then that $f(x)$ must tend to a finite non-negative limit as $x \to \infty$. Clearly enough it is zero. $100/2^{100}$, for instance, is very small. $101/2^{101}$ is nearly twice as small again. This point, however, we shall deal with more formally later.

The graph of $x(\frac{1}{2})^x$ touches that of x, since $(\frac{1}{2})^x$ is just greater than or less than 1 according as x is just less than or greater than 0.

Exercise A

1. Use the rules developed in Section 1.2 to write down the derivates of

$$9^x,\quad (1/5)^x,\quad (16)^x,\quad (1/27)^x,\quad 10^x,\quad (3/5)^x,\quad 60^x.$$

Sketch the graphs of these functions of x.

2. Express the following in the form 2^{Kx} or 3^{Kx} as appropriate:

$$4^x,\quad (1/3)^x,\quad 27^x,\quad (1/16)^x,\quad (\sqrt{2})^x,\quad (0{\cdot}0625)^x.$$

Hence express them as the product of two functions (as in the discussion of 8^x in Section 1.3) and differentiate them.

3. Prove $125^x = (5^x)^3$ and hence differentiate it, using the rule for the product of two functions.

4. Find the derivates of 9^x, $\frac{1}{4}^x$, and $(4/3)^x$ at $x = 5$ and at $x = -5$. Check your answer by sketching their graphs.

5. Find the derivatives of:

$$x^2 2^x,\quad (6^x)/x,\quad x/(5^x),\quad 2^{(x^2)},\quad 6^{1/x},\quad 2^{(2^x)},\quad \sqrt{(12^x)}.$$

6. Sketch the graphs of:

$$2^x - 8k_2 x,\quad x3^{-x},\quad (5^{2x})/x^2,\quad 4^x - \tfrac{1}{4}^x,\quad 4^x - 8^x.$$

7. How many solutions has $2^x = 1{\cdot}5x$?

8. Considering the graph of $3^x/x$, discuss the solubility of $3^x = ax$ for different ranges of values of a.

9. In the same figure sketch the graphs of:

$$x(1/9)^x,\quad x(1/3)^x,\quad x,\quad x3^x,\quad x9^x.$$

2. EXPONENTIAL CHANGE

In this section we shall consider some of the simpler instances of exponential functions that crop up in nature.

2.1 Population growth. Let us consider in what way we would expect a population—of anything: birds or butterflies or bees—to change with time.

Naturally, if we are observing a large enough sample, it seems likely that the births and deaths in a fixed period—say a year—would be roughly proportional to the number in the sample. In a sample of 1000 let us suppose we get, on average, 300 births and 175 deaths a year. Then we get a rise of population of 125 per 1000 in a year: a rise, that is, of one eighth.

This of course would only be *on average*, but let us fix it as a kind of law and see how the idealized population (or 'model' of a population) would grow if it always observed the law.

Suppose the population is originally N. After a year it will have risen by an eighth, to nine-eighths of its original size. Thus it will be $\frac{9}{8}N$. But in the second year it will again grow to nine-eights of what it was at the beginning of that year: that is, to $\frac{9}{8}(\frac{9}{8})\,N = (\frac{9}{8})^2\,N$. After three years it will be $(\frac{9}{8})^3N$. And, in general, if after T years the population is $f(T)$,

$$f(T) = (\tfrac{9}{8})^T\,N.$$

This formula has only been calculated for integral T. It has however a wider significance—if, that is, we suppose the rate of growth of population is regular and smooth and not liable to seasonal fluctuations.

Consider for instance $T = \frac{1}{4}$. We do not know to what fraction of its size the population grows in $\frac{1}{4}$ year; but let it be k. Thus after $\frac{1}{4}$ year we get $f(\frac{1}{4}) = kN$. After two quarters of a year, or $\frac{1}{2}$ year, we get $f(\frac{1}{2}) = k^2N$. Likewise $f(\frac{3}{4}) = k^3N$, and $f(1) = k^4N$. But we know $f(1) = \frac{9}{8}N$. Hence $k^4 = \frac{9}{8}$, and $k = (\frac{9}{8})^{\frac{1}{4}}$. Thus

$$f(\tfrac{1}{4}) = (\tfrac{9}{8})^{\frac{1}{4}}\,N.$$

In the same way the formula can be extended to any rational T.

This implies, incidentally, that in *any* interval of 1 year the population increases by $\frac{1}{8}$. For $f(T+1) = (\frac{9}{8})^{T+1}\,N = \frac{9}{8}(\frac{9}{8})^T\,N = \frac{9}{8}f(T)$.

$$f(T) = (\tfrac{9}{8})^T\,N.$$

Differentiating this we get

$$f'(T) = 0{\cdot}118(\tfrac{9}{8})^T\,N,$$

since $k_{\frac{9}{8}} = 0{\cdot}118$ (to three decimal places). We can interpret this by saying that the rate of growth of population is 0·118 times its momentary size, 0·118 per year per unit in the population.

But we have already seen that the population grows by $\frac{1}{8}$ in a year; so that we might say the rate of growth was $0{\cdot}125f(T)$ or 0·125 per year per unit in the population.

Why the discrepancy between these results?

The answer lies in the fact that we are using two different definitions of the 'growth rate'. The first was the proper calculus definition: the rate at which the population was growing at an instant. But the second was the *average rate of growth calculated over a year.*

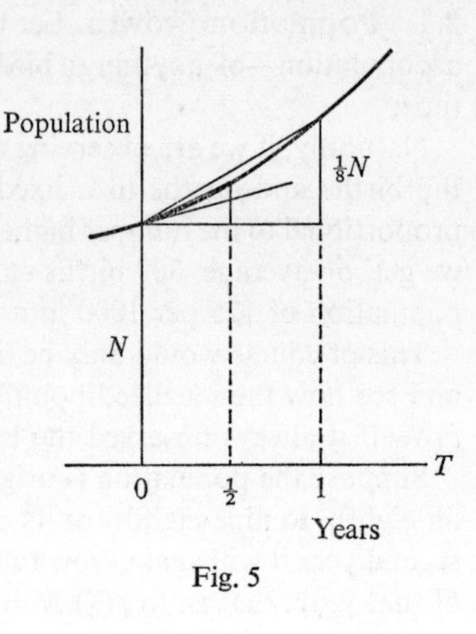

Fig. 5

The difference should be clear in Figure 5. In one year (no matter from where we start) the population grows by one eighth of its original size, N; hence the average rate of growth is $\frac{1}{8}N$ per year. From the figure we see that the actual rate of growth (the gradient of the tangent) must be less than this.

Equally we could work out the average growth rate over a different interval: say half a year. In any half year the population grows from N' (say) to $N'(\frac{9}{8})^{\frac{1}{2}}$. Hence the average rate of change is:

$$[N'(\tfrac{9}{8})^{\frac{1}{2}} - N']/(\tfrac{1}{2})$$

which comes to $0{\cdot}121N'$ per year. Hence the average growth rate is 0·121 per year per unit of population.

As we would expect, this lies between the actual growth rate and the average calculated over a year.

It is also of interest to consider the linear approximation to the size of population near $T = 0$.

$$\begin{aligned} f(\epsilon) &= f(0) + f'(0)\,\epsilon + s_\epsilon \\ &= N + 0{\cdot}118N\epsilon + s_\epsilon. \end{aligned}$$

This of course was a formula worked out for an idealized population, which we expect to give a good approximation to actual population rise in a short time *if N is large enough.*

For instance, in a population of 10000, the increase in a month calculated from this formula would be approximately (neglecting the terms s_ϵ)

$$f(\tfrac{1}{12}) - 10000 = 0{\cdot}118 \times 10000 \times \tfrac{1}{12} \approx 98.$$

This might well be a reasonable estimate.

Let us however apply the formula to a small population ($N = 100$), and see what, if anything, the answer would mean. Here

$$f(\tfrac{1}{12}) - 100 = 0{\cdot}118 \times 100 \times \tfrac{1}{12} \approx 0{\cdot}98.$$

This does not of course mean that in any given colony of 100 there is bound to be an increase of 1 (much less 0·98) in a month. The actual increases in a substantial number of such colonies might fluctuate widely.

However, if we take a large number of these small colonies, we might find that 0·98 was a good approximation *to the average of their increases* in a month.

2.2 Radioactive decay. This is another instance of exponential change, but of exponential decay. It is in fact an instance of a population growth in which we have deaths but no births.

We may think of radioactive material as containing a very large number of unstable atoms. Each of these can change spontaneously into stable ones, while emitting radiation; and in a fixed interval of time there will be a fixed chance that any particular atom changes. The chance is the same for all atoms.

We shall establish a connection between this change and time, and derive a formula for the number of atoms that remain unchanged: that is, for the quantity of matter which is still radioactive, which we shall express as a function of time.

Let us suppose that originally there are N radioactive atoms, and that the chance that any atom changes in 1 minute is 1/100.

After 1 minute, roughly $N/100$ atoms have changed, leaving $0{\cdot}99N$ unchanged. (N is so large for appreciable quantities of matter that this leads to no noticeable error.) Similarly after t minutes there remain $N(0{\cdot}99)^t$ atoms unchanged. Thus if the quantity (let us say the *mass*) of still radioactive matter after t minutes is $f(t)$, and there are N' atoms per unit mass,

$$f(t) = \frac{N}{N'}(0{\cdot}99)^t = m(0{\cdot}99)^t,$$

where originally there were m units of radioactive material.

In this exponential function the constant 0·99 is less than 1. Hence the graph has the dwindling form shown in Figure 6.

We call this 'radioactive decay'. A characteristic of radioactive, or any exponential decay is that of the 'half-life'. Let us calculate the time in which the quantity of radioactive matter is halved: that is, find t where

$$f(t) = m(0{\cdot}99)^t = \tfrac{1}{2}m.$$

Dividing by m and taking logarithms, we get $t \approx 73$.

Thus the quantity of radioactive matter is halved in about 73 minutes.

Now clearly this result is independent of m or the quantity of matter with which we start. We conclude that if from any point we increase t by about 73, the quantity of matter will be halved. This is illustrated by the intervals from A to B, A' to B' and A'' to B'' in Figure 6. We call 73 minutes the *half-life* of the radioactive substance.

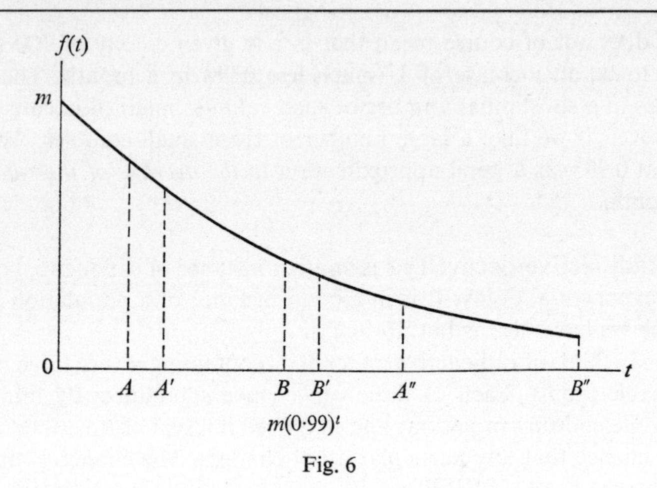

Fig. 6

Of course there is no particular significance in the factor $\frac{1}{2}$: we could equally have worked out the 'third-life' (the time in which it is reduced to a third). But half-life is a useful measure of the rate of decay. The measure can vary widely. For instance the half-life of Polonium (212) is 3×10^{-7} seconds. For Uranium it is $4{\cdot}5\times10^{9}$ years.

Equally the many other cases of exponential decay found in nature will exhibit their characteristic half-lives: for instance, the dying away of a note struck on a piano whose half-life—independent of the original strength of the note—is a measure of the piano's sustaining power.

Example. In three hours a quantity of radioactive substance diminishes from 10 kg to 8 kg. What is its half-life? And what is the chance any atom changes in 1 minute?

Assuming it decays exponentially, we note that in 3 hours it is reduced to $\frac{4}{5}$ths of its original quantity; and deduce that in 6 hours it becomes $10(\frac{4}{5})^2$, in 9 hours $10(\frac{4}{5})^3$; and in general, after t hours

$$f(t) = 10(\tfrac{4}{5})^{\frac{1}{3}t}.$$

To find its half-life we want to know for what value of t

$$f(t) = 10(\tfrac{4}{5})^{\frac{1}{3}t} = 5,$$

or

$$2(\tfrac{4}{5})^{\frac{1}{3}t} = 1.$$

Taking logarithms we get

$$\frac{t}{3}\log\tfrac{4}{5}+\log 2 = 0,$$

from which

$$t = 9{\cdot}32;$$

and therefore its half-life is 9·32 hours.

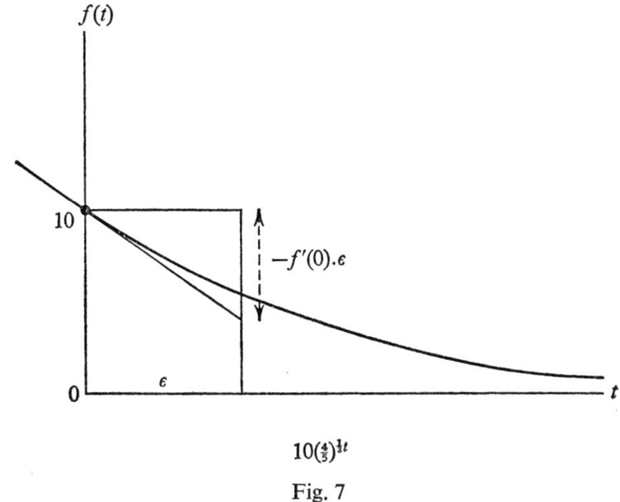

$10(\frac{4}{5})^{\frac{1}{3}t}$

Fig. 7

Secondly, the chance an atom changes in a minute.

In 1 minute, the quantity of radioactive substance is reduced to $10(\frac{4}{5})^{1/180}$. We might work this out, and, comparing it with the original 10 kg, find what fraction of it has changed in a minute, and hence the chance that any atom changed.

However, since the time is small, we can estimate the change by linear approximation near $t = 0$. This is illustrated in Figure 7.

$$f(t) = 10(\tfrac{4}{5})^{\frac{1}{3}t}.$$

Hence $f'(t) = 10[-0{\cdot}223 \times (\frac{4}{5})^{\frac{1}{3}t} \times \frac{1}{3}]$ since $k_{\frac{4}{5}} \approx -0{\cdot}223$.

It follows that

$$\begin{aligned} f(\epsilon) &= f(0) + f'(0) . \epsilon + s_\epsilon \\ &\approx 10 - 10 \times 0{\cdot}223 \times \tfrac{1}{180}, \end{aligned}$$

since the time is 1 minute, or $\epsilon = \frac{1}{60}$. Hence the quantity of radioactive substance diminishes by about 0·0124 kg; or out of 10 kg, about 0·0124 kg of the substance has changed.

It follows that the chance that one atom changes is about 0·00124.

Exercise B

1. Express the relevant physical quantities as exponential functions of t (or T) in each of the following:

(i) a population which increases by $\frac{1}{2}$ every year, after t years,

(ii) a quantity of radioactive substance which is diminished by $\frac{1}{100}$ every minute, after t minutes,

(iii) the value of the Svenndorffi dollar which decreases by $\frac{1}{5}$ every 8 years, after t years (consider the number of 8 year intervals in t years),

(iv) as for (iii), after T months,

(v) the number of bubbles in the sink, where the chance that any one bubble bursts in a second is $\frac{1}{5}$, after t seconds and again after T minutes,

(vi) a crowd around a Hyde Park orator which increases by $\frac{1}{6}$ every 4 minutes, after t seconds.

2. (i) A forest fire doubles its area every 5 hours. By what fraction does it increase in 2 hours?

(ii) In the early stages of a plague, the numbers of the plague-stricken is observed to increase by $\frac{1}{6}$ every week. How long is it before their numbers are doubled?

(iii) In a certain industry the working week in 1945 was 48 hours; in 1960 it was 44 hours. Assuming the working week enjoys an exponential decay, estimate the length of the working week in 1984.

3. In living matter the proportion of carbon which is radioactive is supposed not to vary with time, but it decays from the time of death with a half-life of 5600 years. Date a piece of wood in which the amount of radioactive carbon is found on analysis to be 0·78 of that in a similar specimen of living wood.

4. What is the growth rate of a population which doubles itself every 75 years? By what fraction does it increase in 20 years? (See Section 2.1.)

5. A population increases by a tenth every ten years. How long does it take to double itself? By considering the linear approximation to the population at a particular time, find the expected number by which a colony of 100 inside the population will be increased in six months.

(The derivative of $(1{\cdot}1)^t$ is $0{\cdot}0953\,(1{\cdot}1)^t$.)

6. A population of moths increases by a quarter every three weeks. Using linear approximation, estimate the fraction by which it increases in a day.

7. The chance that a radioactive atom changes in the course of an hour is $\frac{1}{10}$. What is the half-life of the substance? Find the linear approximation near $t = 0$ to the number of atoms unchanged after t minutes, and hence estimate the chance that a particular atom changes in t minutes.

(The derivative of $0{\cdot}9^t$ is $-0{\cdot}105\,(0{\cdot}9)^t$.)

8. If the half-life of a radioactive substance is 1 hour, what approximately is the chance that a particular atom changes in 1 second?

2.3 Probability.

Exponential functions often crop up in connection with probability. Let us take, as a simple example, the radioactive substance discussed in Section 2.2, and consider the life of a particular atom.

What is the probability that it has changed in t minutes?

We can approach this best by considering the probability that it *survives*, or doesn't change for t minutes.

The chance it changes in a minute in $\frac{1}{100}$. Hence the probability it survives is 0·99.

The probability it survives for 2 minutes = (probability it survives for the first minute) × (probability it survives the second minute), which is

$$0{\cdot}99 \times 0{\cdot}99 = (0{\cdot}99)^2.$$

Likewise the probability it survives for 3 minutes is $(0{\cdot}99)^3$, and for t minutes, $(0{\cdot}99)^t$.

But the probability it changes sometime during those t minutes is the probability *that it hasn't survived that time.*

Hence if $f(t)$ is the chance it changes in t minutes,

$$f(t) = 1-(0{\cdot}99)^t.$$

This is the function whose graph we show in Figure 8. We obtain it by sketching the graph of $(0{\cdot}99)^t$ (the dotted line) and subtracting it from 1. As we should expect, $f(t) \to 1$ as $t \to \infty$, for the chance that a particular atom has changed approaches certainty as time increases. Likewise $f(t)$ steadily increases with time: the longer the time, the more likely that the atom has changed.

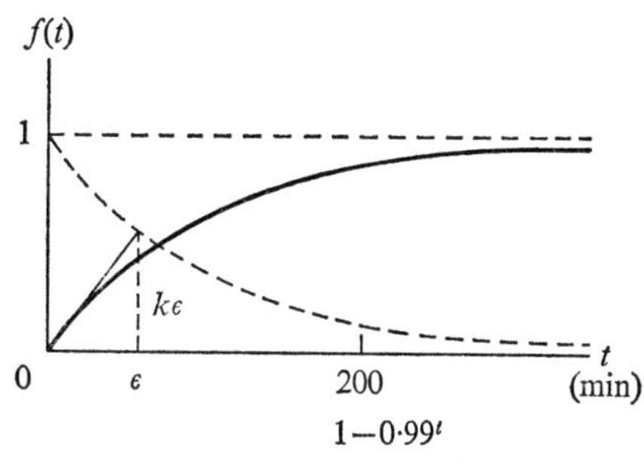

Fig. 8

However $f(0)$ is zero.

$f(t)$ is the chance that the atom changes some time during an interval of t minutes. $f(0)$, then, we interpret as the chance that the atom changes *at the instant* $t = 0$; and it is 0.

This is at first sight puzzling; for, we may argue, the atom has to change at some point of time; how then can the chance it changes at such a point be zero?

But what do we mean by saying an atom changes at a given instant? Surely that *as nearly as we can judge* it changes then: between for instance $-0{\cdot}999$ and $0{\cdot}001$ if we can detect events like this to the nearest thousandth of a minute. But clearly the chance an atom changes in this interval (though small) is not equal to zero. Therefore the atom can change 'at $t = 0$'.

We can see this in another way by taking the linear approximation to $f(t)$ near $t = 0$.

$$f'(t) = 0{\cdot}0101 \times (0{\cdot}99)^t,$$

since $k_{0{\cdot}99} = -0{\cdot}0101$ (to four decimal places). Hence

$$\begin{aligned} f(\epsilon) &= f(0)+f'(0)\,\epsilon+s_\epsilon \\ &= 0{\cdot}0101\epsilon+s_\epsilon. \end{aligned}$$

This result (of a kind which is important in the theory of probability) shows what happens to $f(t)$ as $t \to 0$. Clearly it tends to zero. But for a small interval of time it will not be zero.

Example. In an industrial plant, 40 furnaces are working at a particular time. For each furnace there is a fixed chance p that it will break down in the next 24 hours. After what time has the probability that just one furnace has broken down reached its highest value?

Here we must calculate the probability that just one furnace breaks down in t days—at some time during the t days, not exactly at the end of it, for which the probability would be zero—and find the value of t for which this probability is a maximum. Let us first consider the chance that a particular furnace breaks down in t days; this is the chance that it survives for t days, subtracted from 1: that is

$$1-(1-p)^t \quad \text{or} \quad 1-q^t \quad \text{where } q \text{ is } 1-p.$$

The chance that just one furnace breaks down in t days is:

$40 \times$ (chance a particular furnace breaks down) $\times$ (chance the rest survive);

we multiply by 40, since any one of the furnaces may break down while the rest survive, and we must add the probabilities that these independent events occur. Hence the chance one furnace breaks down is

$$f(t) = 40(1-q^t)\, q^{39t},$$

since the chance 39 survive is $(q^t)^{39}$.

Let us write it

$$f(t) = 40(q^{39t} - q^{40t}).$$

Taking the constant associated with q to be $-k$ (negative since q is less than 1),

$$f'(t) = 40(-39kq^{39t} + 40kq^{40t}),$$

so that $f'(t) = 0$ when $q^t = \frac{39}{40}$.

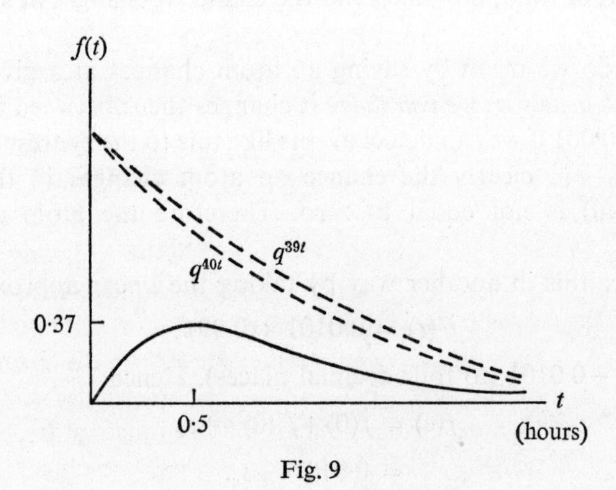

Fig. 9

A sketch of the graph of f shows us that this must be a maximum. The value of t which gives maximum probability therefore can be calculated from a knowledge of q. If p is $\frac{1}{20}$, q is $\frac{19}{20}$, so that $t \log \frac{19}{20} = \log \frac{39}{40}$, whence $t \approx 0{\cdot}49$, or the most likely length of time is just under 12 hours.

It is interesting to note that the answer does not depend on a knowledge of k, and indeed that the maximum probability itself does not depend even on q. It is $f(t) = 40((\frac{39}{40})^{39} - (\frac{39}{40})^{40})$, since q^t is $\frac{39}{40}$. It is also significant that the following argument does not give the right answer: the chance that a furnace breaks down in 24 hours is $\frac{1}{20}$; therefore it is most likely that 2 of the 40 furnaces will break down in 24 hours; therefore the most likely time for just one furnace to have broken down is 12 hours. The reader is left to find for himself the fallacy leading to this conclusion.

Exercise C

1. A burglar's chance of being arrested in a particular month is $\frac{1}{15}$. What is his chance of being arrested in x weeks? Use approximate methods to find his chance of being arrested on a particular day. (The derivative of $(\frac{14}{15})^x$ is $-0{\cdot}069 \times (\frac{14}{15})^x$.)

2. The chance of throwing a six with a weighted die is $\frac{3}{20}$. Find the chance of throwing just one six in N throws. Regarding this as a function of a continuous variable, sketch its graph, and hence estimate the most likely number of throws in which to score just one six. Verify your guess by comparing (without necessarily calculating) the actual probabilities. (The derivative of $(\frac{17}{20})^x$ is $-0{\cdot}16 \times (\frac{17}{20})^x$.)

3. In a certain war, during periods of heavy fighting, a soldier's chance of being put out of action during twelve hours at the front is $\frac{1}{30}$. Find his chance of surviving for t hours, and of becoming a casualty in t hours; find also the chance that there is just one casualty in a regiment of 500 in t hours, and hence calculate the most likely length of time for there to be just one casualty.

4. Of ten pots on the fire, the chance that any particular pot boils over in 1 minute is $\frac{1}{5}$. What is the chance that a particular pot boils over in t minutes? Hence find the chance that just one of the 10 pots boils over in t minutes, and find the most likely time for just one pot to have boiled over.

Do you think the most likely time for two pots to have boiled over is more than, less than, or equal to twice this time? Check your answer by finding the chance that 2 pots boil over in t minutes. (Take the derivative of $(\frac{4}{5})^x$ to be $-0{\cdot}223 \times (\frac{4}{5})^x$.)

3. *e*

3.1 Definition of *e*. We found in Section 1.3 that, for any a, a^x could be expressed in the form 2^{kx}. There is some advantage in doing this. It simplifies algebra to refer everything to a common base; but 2 is not the only base we could use. Equally we could express a^x in the form 3^{kx} or 5^{kx}. The question arises therefore, what number should we use in this way?

Our aim in all this must be simplification. Obviously a great simplification could be achieved by finding a base, a say, whose constant k_a was 1.

We have seen that, approximately, $k_2 = 0{\cdot}7$, and $k_3 = 1{\cdot}1$; therefore we would expect that between 2 and 3 such a number exists. That it does can be seen from the following argument.

If

$$f(x) = 2^{Kx}, \quad f'(x) = k_2 2^{Kx} K = k_2 K 2^{Kx}.$$

Put $K = 1/k_2$. Then we see that

$$f(x) = 2^{x/k_2} \Rightarrow f'(x) = k_2(1/k_2)\, 2^{x/k_2} = 2^{x/k_2}.$$

But

$$2^{x/k_2} = (2^{1/k_2})^x.$$

Put $2^{1/k_2} = a$, and we see that

$$f(x) = a^x \Rightarrow f'(x) = a^x.$$

Here then is our number whose associated constant is 1.

$$2^{1/k_2} \approx 2^{1/0{\cdot}7} = 2{\cdot}7 \quad \text{(to one decimal place)},$$

verifying our guess that it lies between 2 and 3.

More accurately it can be calculated, though not yet, as 2·718; though in fact it is an irrational number. It is a mathematical constant, as is for instance π. Like π it is of fundamental importance in mathematics.

It is the mathematical constant we call e.

3.2 e^x and e^{kx}. We start from the result that

$$f(x) = e^x \Rightarrow f'(x) = e^x.$$

It follows that if

$$F(x) = e^{kx}, \quad F'(x) = e^{kx}k = ke^{kx}.$$

The derivatives therefore of e^{3x}, e^{2x}, $e^{\frac{1}{2}x}$, $e^{-\frac{1}{2}x}$, e^{-2x}, e^{-3x} are

$$3e^{3x}, \quad 2e^{2x}, \quad \tfrac{1}{2}e^{\frac{1}{2}x}, \quad -\tfrac{1}{2}e^{-\frac{1}{2}x}, \quad -2e^{-2x}, \quad -3e^{-3x}.$$

But these themselves are exponential functions. e^{3x}, for instance, is

$$(e^3)^x \approx 20^x; \quad e^{-\frac{1}{2}x} = (e^{-\frac{1}{2}})^x \approx (0{\cdot}6)^x.$$

We can best study the relations between these by sketching their graphs together as in Figure 10.

We see that for $k > 1$, $e^{kx} = (e^k)^x$ tends to ∞ and to 0 more quickly than e^x (since $e^k > e$).

For $0 < k < 1$, it does both these more slowly.

If $k < 0$, $e^k < 1$, and we have an example of exponential decay. Here the derivative is always negative.

The boundary between these categories is $e^{0x} = 1$.

Two functions of x of the form e^{kx} and e^{-kx} (say e^{3x} and e^{-3x}) have graphs symmetrical about the range-axis.

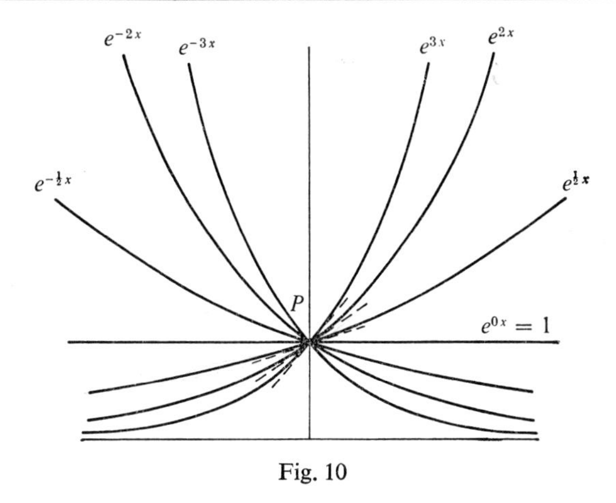

Fig. 10

Lastly, any function of x a^x can be put in this form. For by the definition of logarithms, $a = e^{\log_e a}$. Hence

$$a^x = (e^{\log_e a})^x = e^{(\log_e a)x}.$$

Thus
$$k_a = \log_e a.$$

This verifies, or explains, the rules that we found in Section 1.1. For instance

$$k_6 = \log_e 6 = \log_e 2 + \log_e 3 = k_2 + k_3.$$

In addition $k_a = \log_e a$ will be greater than or less than 0 according as a is greater than or less than 1.

In future we shall always write exponential functions in the form e^{kx}, noting that we have to deal with a case of exponential growth if $k > 0$, and of decay if $k < 0$: though for negative constant, we usually write the function e^{-kx}, preserving the convention that k itself is positive.

Exercise D

1. Express the following functions in the form e^{kx}; write down their derivatives, and sketch their graphs:

$$(e^5)^x, \quad (e^x)^2, \quad 1/e^x, \quad (\sqrt{e})^x, \quad e^{2x}/e^{5x}, \quad (1/e^{3x})^4,$$
$$(e^x)^3\,(e^{-x})^7, \quad 2^x, \quad (0{\cdot}1)^x, \quad 10^x, \quad \sqrt{[(\tfrac{3}{4})^x]}.$$

2. By writing e^{kx} as $P_k(e^x)$, prove its derivative is ke^{kx}.

3. If $E(x) = e^x$, what are the functions EP_2 and P_2E? Use the product rule to differentiate $e^{(x^2)}$ and sketch its graph.

4. Express the following as product functions and hence differentiate them:

$$e^{1/x}, \quad e^{1-2x}, \quad (1+e^{-x})^2, \quad (1-e^{2x})^{-1}, \quad e^{(e^x)}.$$

3.3 Behaviour of e^x. So far we have discussed the general behaviour of such functions of x as 2^x, $\frac{1}{4}^x$ and 3^{3x}. Let us now examine it a little more closely as, in particular, x tends to ∞ or $-\infty$. We know of course that $e^x \to \infty$ as $x \to \infty$, and that $e^x \to 0$ as $x \to -\infty$; but—putting it rather loosely—to *how big* an infinity and to *how small* a zero?

We have as yet no criterion for gauging how big is an infinity; but at least we can compare two infinities. For instance, if we compare the graphs of x^2 and x^3 we can see that x^3 tends to a larger infinity than x^2 as $x \to \infty$. Likewise x^3 tends to a smaller zero than x^2 as $x \to 0$.

How then would e^x compare with x or x^2 or x^3 as $x \to \infty$?

The answer seems to lie in the behaviour of such functions of x as e^x/x and e^x/x^2 and e^x/x^3.

Take the first, for instance: e^x/x or $x^{-1}e^x$.

If it tends to ∞ as $x \to \infty$, then we will say that e^x tends to a larger infinity than x. If to 0, then that e^x tends to a smaller infinity. If to a finite non-zero limit, we will say that the infinities are of the same order of magnitude.

Or suppose we are considering to how small a zero e^x tends as $x \to -\infty$.

We could compare it, for instance, to x^{-3}, which also tends to zero. If $e^x/x^{-3} = x^3e^x$ tends to zero as $x \to -\infty$, we say that e^x tends to the smaller zero. And so on.

The last point can be seen from a slightly different point of view. If we consider the product x^3e^x, as $x \to -\infty$, one part of it, x^3, tends to $-\infty$; the other, e^x, tends to zero. If then the product tends to zero, we can say the zero associated with e^x is, so to speak, more powerful than the infinity associated with x^3.

Let us start by comparing the infinities associated with e^x and x^3.

Let

$$f(x) = e^x/x^3 = x^{-3}e^x.$$

Clearly $f(x) \to \infty$ as $x \to 0$, and tends to 0 as $x \to -\infty$.

$$f'(x) = x^{-3}e^x - 3x^{-4}e^x$$
$$= x^{-4}e^x(x-3).$$

So that $f'(x) \gtreqless 0$ according as $x \gtreqless 3$.

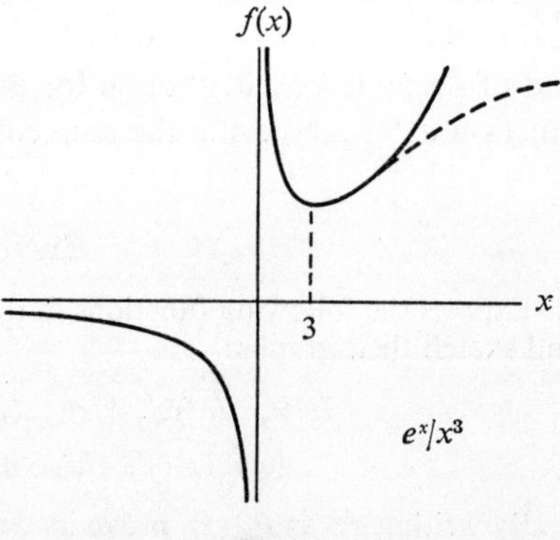

Fig. 11

The function therefore has a minimum where $x = 3$, and thereafter increase as x increases. This would be consistent with $f(x)$ tending to a finite limit as in the dotted alternative shown in Figure 11, or to ∞, as in the unbroken alternative.

Let us with the help of this result consider the behaviour of e^x/x^2. Now $e^x/x^2 = x^{-2}e^x = (x^{-3}e^x)\,x$. But we know that $x^{-3}e^x$ tends to a finite limit or to ∞ as $x \to \infty$. In either of these cases we see that $x^{-2}e^x$ must tend to ∞. Hence e^x tends to a larger infinity than x^2.

As a matter of fact we can prove in a similar way (by proving, say, that $x^{-4}e^x$ tends to a finite limit or ∞) that $x^{-3}e^x$ does tend to ∞. We can prove that e^x/x^n tends to ∞ as $x \to \infty$ for *all* positive n. From which we conclude that e^x tends to a more powerful infinity than *any* power of x as $x \to \infty$.

We could use this result to prove directly that as x tends to ∞, e^{-x} has a smaller zero than any (negative) power of x. It is instructive however to consider the problem by examining independently the function

$$e^{-x}/x^{-1} = xe^{-x}.$$

If this tends to zero as $x \to \infty$, then we will conclude that the zero represented by e^{-x} is more powerful than the infinity represented by x, or smaller than the zero represented by x^{-1}.

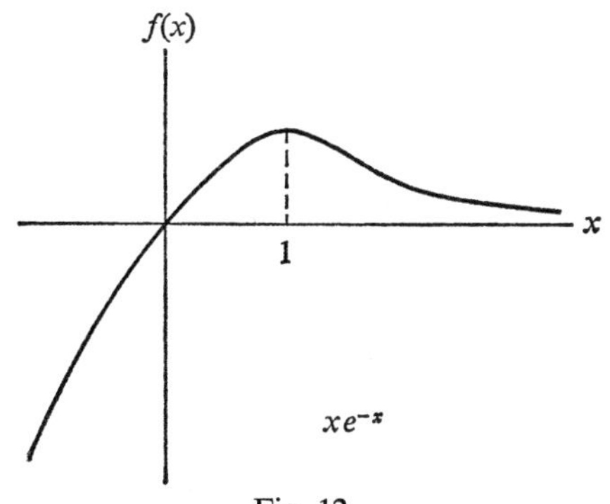

Fig. 12

Here $f(x) = xe^{-x} \gtreqless 0$ according as $x \gtreqless 0$. Clearly as $x \to -\infty, f(x) \to -\infty$.

$$f'(x) = e^{-x} - xe^{-x} = e^{-x}(1-x),$$

so that $f'(x) \lesseqgtr 0$ according as $x \gtreqless 1$.

We see then that $f(x)$ has a maximum where $x = 1$, and thereafter decreases while remaining positive.

This would be consistent with it tending to a non-negative finite limit L. We have to determine whether L is positive or zero.

However if we consider x^2e^{-x} we can show in the same way that it tends to a non-negative limit L' as $x \to \infty$. An argument similar to that used when discussing $x^{-2}e^x$ will prove that L is zero. The reader is left to work out the details of this for himself.

We can sum all this up by saying that e^x and e^{-x} tend to extremely powerful infinities and extremely powerful zeros: more powerful than those associated with any power of x.

Thus, for instance, while as $x \to \infty$ both e^x and x^7 tend to ∞, their quotient $e^x/x^7 \to \infty$ since e^x has the bigger infinity.

Or seen in another way, while as $x \to \infty$ $e^x \to \infty$ and $x^{-7} \to 0$, their product $x^{-7}e^x \to \infty$.

Likewise while as $x \to -\infty$ $e^x \to 0$ and $x^5 \to -\infty$, their product $x^5 e^x$ tends to zero.

Exercise E

1. Sketch the graphs of

$$\text{(i) } e^x + e^{-x}, \quad \text{(ii) } e^x - e^{-x} \quad \text{and} \quad \text{(iii) } \frac{e^x - e^{-x}}{e^x + e^{-x}}.$$

2. Sketch the graphs of $e^{2x} - 3e^{3x}$, $2x - e^{-x}$, $2x + e^{-x}$, $(x-1)\,e^x$.

3. Sketch the graphs of x^2e^{3x}, $x^{-2}e^{-x}$, x^3e^{-2x}.

4. If $f(x) = e^x/x$, simplify $[f(2x)/f(x)]$ and deduce the behaviour of $f(x)$ as x tends to infinity. What can you say about xe^x as x tends to $-\infty$? Discuss the behaviour of $x^{-2}e^x$ and of $x^{-10}e^x$ as x tends to ∞.

5. Find the gradient of the line through the origin which touches the graph of e^x. For what values of k has the equation $e^x = kx$ (i) two roots, and (ii) no roots?

6. By considering the function e^x/x determine the number of roots of the equation $e^x = 3x$.

7. How many roots has the equation $e^{2x} = 10x^2$? Discuss the solubility of the equation $e^{2x} = ax^3$ for various ranges of the values of a.

8. Of a piece of radioactive material the number of grams which are still radioactive after t years is given by the formula $A(t) = Ke^{-kt}$. Find the value of k if the half-life of the material is 10 years.

9.** The problem is to form a probability model for the recurrence of a completely random event, like a bell in a bell-tower clanging owing to the action of the wind; or the distribution of these events along a time scale. We will assume that one such event is no more likely to occur between two points of time a length t units apart, than between any other two points of time a length t units apart: this being true for all t. Thus the probability that just one event occurs in t units of time may be regarded as a function of t, $p_1(t)$ say. Making reasonable assumptions (which you should state) show that $p_1(\epsilon) = P\epsilon + s_\epsilon$, for some constant P.

The chance that r events occur in t units of time we will write $p_r(t)$. The problem is to find $p_0, p_1, p_2, \ldots$.

If the chance no event occurs in 1 unit of time is q, prove that $p_0(t) = e^{-Qt}$, where $Q = -\log q$ so that $q = e^{-Q}$.

Given a length of time t units, break it up into t/n equal lengths of n units. Calculate in terms of t, n, P and Q the probability that just one event occurs in

t units of time. Hence, by considering the limit as $n \to 0$, prove that $p_1(t) = Pte^{-Qt}$. Sketch the graph of this and of $p_0(t)$ in the same figure, calculating the length of time for which the chance that just one event has occurred has reached its greatest value. For this length of time, is it more likely that 1 event has occurred or that no event has occurred?

Considering the linear approximations to $p_0(t)$ and $p_1(t)$ near the origin or otherwise, produce plausible reasons for supposing that $P = Q$. Assume this is true from now on.

Prove that $p_2(t) = \frac{1}{2}p^2t^2e^{-Pt}$, by a method similar to that used for $p_1(t)$. Sketch its graph, and find the time in which the chance that just two events have occurred has risen to its highest value.

Repeat for $p_3(t)$. (You should get $\frac{1}{6}P^3t^3e^{-Pt}$.)

What, in this model, is the average distribution of events per time? Suggest a way in which, from listening to the bell-tower for some time, you could estimate these probabilities.

Distributions of this kind are called Poisson Distributions.

(*a*) Assuming that the arrival of visitors at a particular house forms a Poisson Distribution during daylight hours, and that the average rate of arrival is 5 per six hours, what is the chance the mistress of the house will be able to sleep from 2 p.m. to 4 p.m. without being disturbed, and what is her chance of only being woken up once?

(*b*) The chance of meeting one or more police cars on any one mile of a certain road is $\frac{1}{20}$. What is the most likely number of police cars for one to meet on 60 miles of the road? What are the chances of meeting just this number, of meeting one more, and of meeting one less?

(*c*) On a long thin strip of chromium plating, graduated in millimetres there occur isolated spots of rust. If the strip is 10 metres long, and it is observed that within 540 of the graduations there occur no spots of rust, in how many would one expect there to be just one spot of rust, and how many spots of rust are there most likely to be altogether?

8

THE LOGARITHM FUNCTION

Closely associated with the exponential function is, as we have seen, the logarithm function. It is the purpose of this chapter to study its uses and behaviour. Much of the work we do will be true (with modification) of any logarithm function: of $\log_2 x$ and $\log_{10} x$ as much as of $\log_e x$. $\text{Log}_e x$ is, however, of such dominant importance in mathematics that we shall be concerned with it almost entirely. Logarithms to the base e are called 'natural logarithms'; and when we do not specify a base, as in $\log x$, it will be assumed that we are dealing with natural logarithms. The function then with which we are here concerned is f, where

$$f(x) = \log x.$$

Other logarithm functions are simply multiples of the natural logarithm function: for

$$\log_a x = \log_a e \log_e x.$$

There is thus no point in giving them a separate discussion.

1. log x

1.1 The graph of log *x*. Log x is defined to be the power to which we raise e to obtain x. The relation between x and its logarithm is therefore most clearly seen from the exponential curve.

Figure 1 shows the exponential curve.

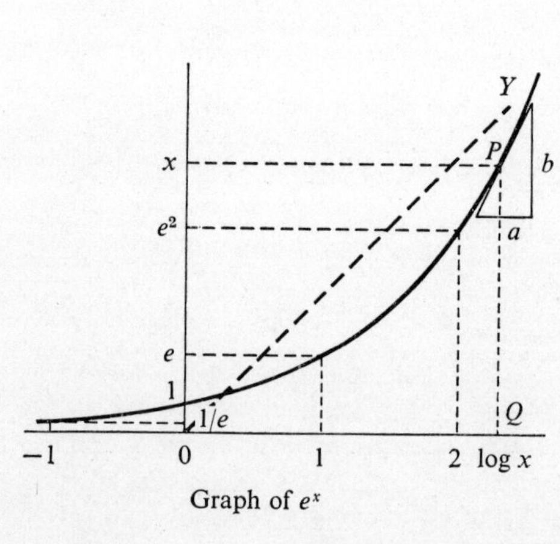

Fig. 1

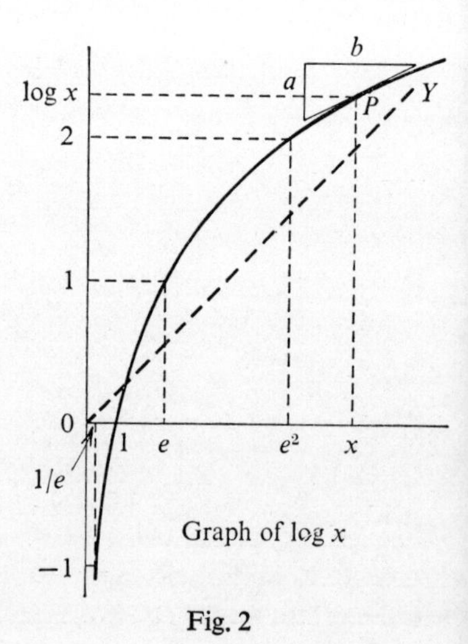

Fig. 2

If we mark x on the range axis, $\log x$ appears in the corresponding position on the domain axis; for $\log x$ is the power to which we raise e to get x. In particular, we see the logarithm of e^2 (see Figure 1) is 2; the logarithm of e is 1; the logarithm of $1/e$ is -1.

It seems then that Figure 1 can be read as a kind of *graph of* $\log x$, with the values of x ranged along the vertical axis, and those of $\log x$ along the horizontal axis. Hence to get the conventional graph of $\log x$ we have only to interchange these two axes: reflect the whole figure, that is, in their angle bisector, OY.

The result of this is shown in Figure 2.

We note that, again, the logarithms of e^2, e, $1/e$ are 2, 1, -1. In particular we see that the logarithm of 1 is 0 (the power to which we raise e to get 1 is 0).

Further we note that the logarithms of numbers less than 1 are negative, and that as $x \to 0$, $\log x \to -\infty$.

As $x \to \infty$, $\log x$ also tends to ∞, but not very emphatically. This seems reasonable, since for large x a large increase in x (say from e^{100} to e^{101}) produces a relatively small increase in $\log x$ (here 1).

For $x < 0$, $\log x$ is undefined.

The relationship between the logarithmic and exponential function is important:

$$\text{if} \quad e^x = y \quad \text{then} \quad x = \log y;$$

$$\text{if} \quad \log x = y \quad \text{then} \quad x = e^y.$$

1.2 Derivative of log x. We can use various methods to differentiate $\log x$. One of the simplest is to compare Figures 1 and 2.

In Figure 2, the gradient of the tangent at P (or the derivative of $\log x$) is a/b. But in Figure 1, a/b is $1/$(gradient at P). The gradient at P is equal to PQ (the fundamental property of the exponential function), or to x.

Hence the derivative of $\log x$ is $1/x$.

In more analytical form, we could argue as follows:

$$f(x) = \log x \Rightarrow e^{f(x)} = x.$$

But if $e^{f(x)} = x$, differentiating both sides, we get

$$e^{f(x)}f'(x) = 1.$$

Hence
$$f'(x) = \frac{1}{e^{f(x)}} = \frac{1}{x}.$$

Or suppose that we have no knowledge of the exponential function, and are dealing simply with a general logarithm, $\log_a x$. We could use the formal properties of logarithms ($\log AB = \log A + \log B$, and so on) to differentiate them very much as we differentiate 2^x in Chapter 7. We leave the details of this to the reader (see Exercise A, Question 4); but we would find that the derivative was k/x, where k was a constant depending on a.

We mention this here to show that the approach actually developed to logarithmic and exponential functions in this and the last chapter, was not the only one possible. Equally we could have started from $\log_a x$, defined e as that number for which the constant k was 1, and built up our theory of exponential functions from that of logarithms. There would be little to choose between that and the method we have adopted.

***1.3 Formal properties of log x.** It is interesting to find that we can establish the formal properties of logarithms from their differential properties.

Consider, for instance, the function

$$F(x) = \log 3x.$$

Differentiating, by the product rule, we get

$$F'(x) = \frac{1}{3x}3.$$

(In general, if

$$F(x) = \log g(x), \quad F'(x) = \log' g(x)\, g'(x) = \frac{1}{g(x)}\, g'(x).)$$

Hence $$F'(x) = \frac{1}{x}.$$

But this is also the derivative of $\log x$. But we know that if two functions have always the same derivative, they differ by a constant.

Hence

$$\log 3x = \log x + C.$$

Put $x = 1$, and we get $\log 3 = C$, since $\log 1$ is zero. Hence

$$\log 3x = \log x + \log 3.$$

Obviously the same method could be used if 3 were replaced by any other number; so that in general

$$\log xy = \log x + \log y.$$

This is a piece of circular argument. We used the law of indices to find the derivative of e^x. Now we have used the properties of e^x to prove something which, lightly disguised, is really the law of indices. But the argument does suggest an alternative and more sophisticated approach to these functions, which becomes important in more advanced analysis.

This would be to start, not from our vague and informal notions of indices and logarithms, *but from the function whose derivative is* $1/x$. (There is a theorem in later calculus which shows that such a function must exist.)

Then we could examine its formal properties, as above, and develop a whole theory of functions of this type: proving incidentally that it has the property which is normally used to define a logarithm.

To do all this would be beyond the scope of the present book; but it is the approach normally adopted in rigorous analysis.

Exercise A

1. Using the formal properties of logarithms ($\log ab = \log a + \log b$, and $\log (a^p) = p \log a$) express the following in terms of $\log x$, and hence differentiate them:

$$\log 5x, \quad \log (x/2), \quad \log x^3, \quad \log (1/x), \quad \log (5/x^2).$$

Also express them in the form $\log f(x)$, for some function f, and hence differentiate them by the product rule, checking your answers against those obtained above.

2. Differentiate the following functions of x:

$$(\log x)^5, \quad \log x^5, \quad x^2 \log x, \quad x \log 4x, \quad \log \sqrt{x}, \quad \log (1-2x),$$

$$\log e^x, \log (x^2-1), \quad 1/(\log x), \quad x/(\log x), \quad \log \frac{x}{1-2x}, \quad e^{\log x}.$$

3. Sketch the graphs of the following:

$$\log (x+3), \quad \log (1-2x), \quad \log (1/x), \quad (\log x)^3, \quad \log (1+x^2), \quad \log (x^2-x-2),$$

having first sketched the graphs of the functions of x in the brackets.

4. $H(x)$ is a function of x with the property that $H(ax) = H(a)+H(x)$ for all a and x. Show that $H(1) = 0$. By differentiating each side of this equation and then putting $x = 1$, show that $H'(a) = k/a$, for some constant k depending on the function.

5. $E(x)$ is a function of x with the property that $E(a+x) = E(a)\,E(x)$ for all a and x. Find $E(0)$. By differentiating each side and putting $x = 0$, prove that $E'(a) = kE(a)$, for some constant k depending on the function. Use this to prove that the derivative of 2^x is $k2^x$, where k is its derivative where $x = 0$. What assumptions have you made in the proof?

***6.** Assuming that $\log_a x$ is differentiable where $x = 1$, write down the linear approximation of $\log_a(1+\epsilon)$. (Do not assume any other differential properties of logarithmic functions.) Hence, using the formal properties of logarithms, and considering $\log (X+\epsilon)-\log X$, form a linear approximation to $\log_a x$, near $x = X$, and deduce that the derivative of $\log_a X$ is k/X for some constant k. What is the constant k?

1.4 Behaviour of log x. It is clear that the derivative of $\log x$, $1/x$, confirms our previous sketch of its graph. As $x \to \infty$, the gradient tends to 0, though $\log x$ itself tends to ∞. As $x \to 0$, the gradient tends to ∞, and $\log x$ tends to $-\infty$.

Let us now turn our attention to the question we answered for e^x: to how big an infinity (or minus infinity) does $\log x$ tend? We will compare them again with the infinities associated with x, x^2, $1/x$ (as $x \to 0$), and so on. Let us consider, as an example, $(\log x)/x^2$ as $x \to \infty$.

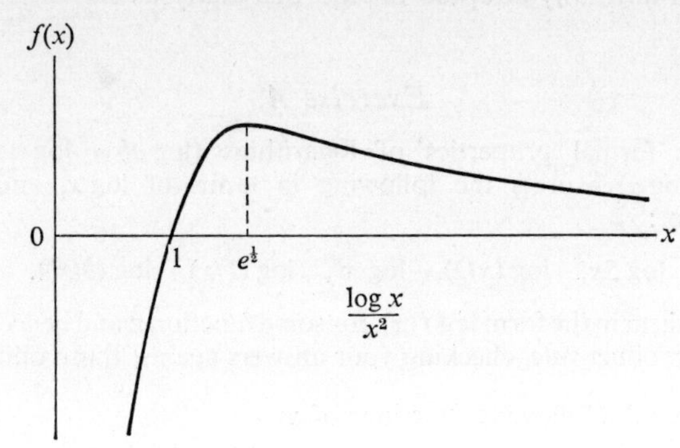

Fig. 3

Evidently if $f(x) = (\log x)/x^2$, $f(x) \to -\infty$ as $x \to 0$.
$f(x) \gtrless 0$ according as $x \gtrless 1$.
$f'(x) = x^{-3} - 2x^{-3} \log x = x^{-3}(1 - 2\log x)$.
But $\log x \gtreqless \frac{1}{2}$ according as $x \gtreqless e^{\frac{1}{2}}$.
Hence $f'(x) \gtreqless 0$ according as $x \lesseqgtr e^{\frac{1}{2}}$.
This shows that $f(x)$ has a maximum at $x = e^{\frac{1}{2}}$, and thereafter decreases while remaining positive. Hence it tends to a non-negative finite limit which may be zero.

But we can show it is zero by considering, in the same way, $(\log)/x$.

We can again show that this tends to a finite limit, L say.

But
$$f(x) = (\log x)/x^2 = \frac{1}{x}(\log x)/x.$$

It follows that since $(\log x)/x$ tends to a finite limit L as $x \to \infty$, $f(x)$ tends to zero.

Thus $\log x$ tends to a weaker infinity than x^2.

We could prove by a similar argument that it is weaker than that of x, and indeed of any positive power of x. (See Exercise B, Question 2.)

As for its infinity as $x \to 0$, we can investigate this by comparing it with that of $1/x$. We would consider the function $(\log x)/(1/x)$, or $x \log x$.

This again we can prove to tend to zero. In fact, just as e^x provided infinities more powerful than any power of x, those of $\log x$ are weaker. Anything of the form $x^a \log x$ (for positive a) tends to zero as x tends to zero: in fact any zero like x^a is capable of swamping the infinity associated with $\log x$.

Exercise B

1. Sketch the graphs of:

$$x^{-2}\log x, \quad (x-1)^2\log x, \quad \log x-2x^2, \quad x^{-1}\log|x|.$$

2. Prove that $x^{-1}\log x$ tends to a finite limit as x tends to infinity. By proving that, likewise, $x^{-\frac{1}{2}}\log x$ tends to such a limit, prove that $x^{-1}\log x \to 0$.

3. Prove, by considering also the limit of $x^{\frac{1}{2}}\log x$, that $x\log x \to 0$ as $x \to 0$.
Sketch the graphs of $x^{\frac{1}{2}}\log x$, $x\log x$, $x^2\log x$, distinguishing carefully their forms at the origin.

4. Sketch the graph of $x^3\log|x|$.

5. Sketch the graph of $x^{1/x}$, $x > 0$, by first sketching the graph of the logarithm of this function. What is the limit of the Nth root of N as $N \to \infty$?

6. Find the number of roots of the equation $3\log x = x$. Sketch the graph of $3\log x - x$ and use Newton's method to find one of the roots to 1 decimal place.

7. Discuss the number of roots of $\log x = a/x^2$ for different ranges of values of a.

8. What are the linear approximations to e^{kx} and to $\log(1+x)$ near $x = 0$?

9. If $f(x) = x^x$ $(x>0)$, prove, by first differentiating the equation

$$\log f(x) = x\log x$$

that $f'(x) = x^x(1+\log x)$. Sketch the graph of f.

10. Use the method of Question 9 to differentiate the functions of x:

$$x^{(x^2)}, \quad x^{\log x}, \quad x^3(1-3x)^{-\frac{1}{2}}/(1+3x^2), \quad (\log x)^x.$$

2. FUNCTION FROM DERIVATIVE

We found, in Chapter 5, how to integrate any power of x, except x^{-1}.

One of the most important aspects of the logarithmic function is that it enables us to fill this gap, for if $f'(x) = 1/x$, *and x is positive*, then

$$f(x) = \log x + C.$$

If x is negative, clearly this is not true, since $\log x$ is not defined for negative x. On the other hand, it seems likely that there is a function, with negative domain, whose derivative is $1/x$. Let us look at the matter graphically.

Figure 4 shows, for positive x, the graph of $\log x$; with lines whose gradients are its derivatives for $x = 1, 2, 3, 4, 5$: that is $1, \frac{1}{2}, \frac{1}{3}, \frac{1}{4}, \frac{1}{5}$.

Now imagine a function with negative domain, whose derivatives for $x = -1, -2, -3, -4, -5$ are $-1, -\frac{1}{2}, -\frac{1}{3}, -\frac{1}{4}, -\frac{1}{5}$.

Evidently its graph could touch the five lines on the left of Figure 4: or could be, in fact, the mirror image of the graph of $\log x$ in the vertical axis.

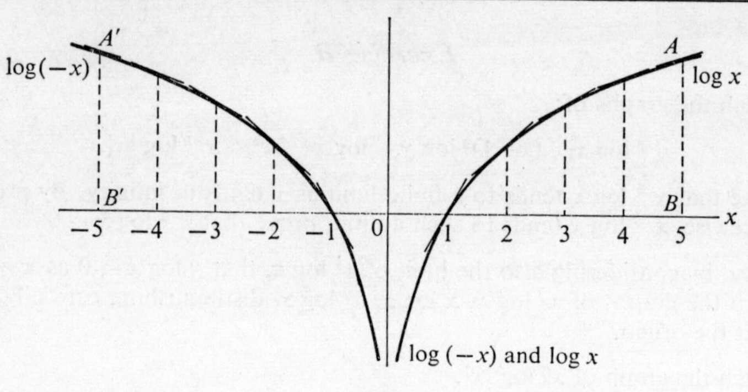

Fig. 4

Let this graph be that of a function $f(x)$. Then, for instance,

$$f(-5) = A'B' = AB = \log 5.$$

Likewise $f(-4) = \log 4$, and in general,

$$f(x) = \log(-x),$$

where $\log(-x)$ *is* defined, since all elements of the domain of f are negative.

Hence, if $f'(x) = 1/x$, *and x is negative*

$$f(x) = \log(-x)+C.$$

We could check this conclusion by differentiating $\log(-x)$. We get

$$f'(x) = \frac{1}{-x}(-1) = \frac{1}{x}.$$

We can unite these two formulae for positive and negative x in the form

$$f'(x) = 1/x \Rightarrow f(x) = \log|x|+C.$$

2.1 Integration. We can put this result to use in simple problems of integration.

The integral of $1/x$ is $\log|x|$. Hence integrals of $3/x$, $-1/x$, for example, are $3\log|x|$, $-\log|x|$.

Or consider the integral of $1/(2-3x)$.

If $2-3x > 0$, $\log(2-3x)$ exists and its derivative is $-3\dfrac{1}{2-3x}$.

If $2-3x < 0$, $\log-(2-3x)$ exists and its derivative is $3\dfrac{1}{-(2-3x)}$ which is the same as $-3\dfrac{1}{2-3x}$.

Hence, for all x (except $x = \frac{2}{3}$), the integral of $\dfrac{1}{2-3x}$ is $-\frac{1}{3}\log|2-3x|$.

Take another example.

If we differentiate either $\log(1-3x^2)$ or $\log -(1-3x^2)$, we get $-6x.1/(1-3x^2)$.

Hence the integral of $x/(1-3x^2)$ is $-\frac{1}{6}\log|1-3x^2|$.

Example. Calculate the work done in compressing a cylinder of gas with constant cross-section from length L_1 to length L_2.

In this problem, we can imagine a piston head moving down the cylinder and compressing the gas. As it moves it will be opposed by force arising from the pressure of gas inside the cylinder, and clearly this pressure will increase as the piston moves further down.

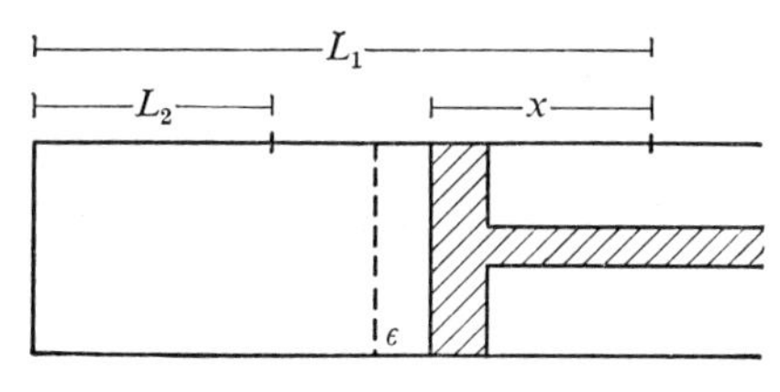

Fig. 5

If we suppose the work done when it has moved a distance x is $f(x)$, then the additional work done in moving a further distance ϵ is $PA\epsilon + s_\epsilon$, where P is the pressure and A the area of cross-section.

Now, by Boyle's law,

$$P = k/\text{volume} = k/A(L_1 - x)$$

since there is a constant quantity of gas in the cylinder.

Hence we have the equation

$$f(x+\epsilon) = f(x) + \frac{Ak}{A(L_1-x)}\epsilon + s_\epsilon.$$

It follows from the definition of linear approximation that

$$f'(x) = \frac{k}{L_1 - x}.$$

Integrating this we have

$$f(x) = -k\log|L_1 - x| + C.$$

We know that $f(0) = 0$; hence

$$C = k\log L_1.$$

But we are trying to find the work done when $x = L_1 - L_2$, that is

$$\begin{aligned} f(L_1-L_2) &= -k\log|L_1-L_1+L_2| + k\log L_1 \\ &= k[\log L_1 - \log L_2] = k\log(L_1/L_2). \end{aligned}$$

The answer is interesting, because it is seen to depend (apart from the nature and quantity of gas, which determine k) only on the *ratio* of L_1 and L_2. This means that if we halved the volume of gas we would be doing a fixed amount of work, $k \log 2$. If we then halved it a second time we would do the same amount, and so on. Clearly then we could never reduce the volume to zero, for this would require an infinite amount of work. We can however (theoretically at least) approach as nearly as we like to this by successive halvings, reducing it to (say) $(\frac{1}{2})^n$ of its original volume by doing $nk \log 2$ units of work. This suggests the increasing difficulty of compression. For each successive halving requires a greater force to accomplish it, in so far as the same amount of work is required in moving just half the previous distance. This is also clear from Boyle's law.

Exercise C

1. Write down the most general expressions for $f(x)$ if $f'(x)$ is:

$$1/5x, \quad 1/(-x), \quad 1/(3x-2), \quad 1/(2-3x), \quad x/(1+x^2), \quad e^x/(1-2e^x).$$

2. $f'(x) = 1/(1+3x)$ and $f(1) = 5$. Find $f(x)$ and specify the domain in which your answer is valid.

If further $f(-2) = 7$, find $f(x)$ for all x (except $x = -\frac{1}{3}$) and sketch its graph.

3. Repeat Question 2 for $f'(x) = 1/(3-x)$ and $f(0) = 0$ and $f(5) = -3$.

4. Find the work done by the gas inside a cylinder in pushing the piston confining it from 1 m from the end to 3 m from the end. The cross-section area of the cylinder is 0·2 m² and the pressure of the gas is k/volume, where $k = 340$ J. [log 3 = 1·0986.]

2.2 Exponential growth. In general we know that the derivative of $\log f(x)$ is $f'(x)/f(x)$. Hence we can argue that an equation of the form

$$g'(x) = f'(x)/f(x)$$

implies

$$g(x) = \log|f(x)| + C.$$

This principle can be used to solve one of the most important equations in mathematics. From Chapter 7 we know that if $f(x) = Ae^{kx}$, then $f'(x) = Ake^{kx} = kf(x)$: that is to say the rate of change of an exponential function is proportional to the function itself.

Now we can prove the converse: that if we *know* that the rate of change of a particular function of x is proportional to itself—and this is a very common condition in applied mathematics— then the function must be of the form Ae^{kx}, where A is any constant, and k the constant of proportionality.

For if
$$f'(x) = kf(x),$$
it follows that
$$\frac{f'(x)}{f(x)} = k.$$

Now taking the functions which when differentiated give us the two sides, we see that
$$\log|f(x)| = kx+C.$$
Hence $|f(x)| = e^{kx+C} = Ae^{kx}$, where $A = e^C$, and is therefore positive. If $f(x)$ is itself positive, this becomes,
$$f(x) = Ae^{kx}.$$
If on the other hand it is negative, it becomes
$$f(x) = -Ae^{kx}.$$
Hence we can say that the *general* solution of the equation is
$$f(x) = Ae^{kx},$$
where the constant A can take either positive or negative values.

We will illustrate the fundamental equation in some examples which show its use in applied mathematics.

Example 1. The numbers in a colony of insects are observed to be growing at the rate 2 insects per day per 50 insects in the colony. How long does the colony take to double itself?

This is an example akin—although seen from the reverse point of view—to those in Chapter 7, Section 2.1. If $f(t)$ is the number of insects in the colony after t days from some origin for time, then we know that
$$f'(t) = 2\frac{f(t)}{50} = 0{\cdot}04f(t).$$

The solution of this equation is of the form $Ae^{0{\cdot}04t}$. A is the 'constant of integration'; it can be found from knowledge of the numbers in the population at a particular time. For instance, if it was known that after two days there were 500 insects, then
$$f(2) = Ae^{0{\cdot}04\times 2} = 500,$$

from which clearly we can calculate A. However the solution of the present problem does not depend on this. We want to find the time in which the population is doubled: in which, to take a particular case, it is increased from A (the value when $t = 0$) to $2A$. This is T, where
$$Ae^{0{\cdot}04T} = 2A,$$
or
$$T = (\log 2)/0{\cdot}04 = 17{\cdot}3.$$

It is evident that we would get this value wherever we started, since we are free to choose our origin of time at any point we like. More generally, when t is increased by the value $(\log 2)/0{\cdot}04$, $f(t)$ grows from

$$Ae^{0\cdot04t} \quad \text{to} \quad Ae^{0\cdot04t+\log 2} = 2Ae^{0\cdot04t}:$$

that is to say, it is exactly doubled in this interval of time of 17·3 days.

Example 2. A heated object is placed in a room where the temperature is maintained at 20 °C. If it takes 5 minutes for its temperature to fall from 70 to 60 °C, how much longer will it be before it reaches 45 °C?

We shall use the law that the rate of cooling is proportional to the difference between the object's temperature and that of its surroundings. (This is known as 'Newton's law of cooling'; the reader might infer it for himself as the simplest law which could control the cooling, where the rate of loss of heat in some way depends, as we would naturally expect it to depend, on the temperature difference.) Exactly how would its temperature vary with time?

It will be convenient to take a new zero to the temperature scale at room temperature, so that initially the temperature was 50° and this fell to 40° after 5 minutes. We are to find how much longer it will be before the temperature is 25°.

On this scale the temperature as a function of the time, t minutes, elapsed since the temperature was 50°, will be denoted by $T(t)$°.

Then

$$T'(t) = -kT(t), \tag{i}$$

where k is a positive constant and $-T'(t)$ is the rate of *loss* of heat.

The general solution of the equation is,

$$T(t) = Ae^{-kt}.$$

Imposing the initial conditions: $T(0) = 50$, then $A = 50$; so the particular solution is

$$T(t) = 50e^{-kt}. \tag{ii}$$

We know that after 5 minutes the temperature is 40°, hence

$$50e^{-5k} = 40.$$

Therefore $$e^{5k} = \tfrac{50}{40} = 1{\cdot}25,$$

or $$k = \tfrac{1}{5}\log 1{\cdot}25.$$

The time when the temperature is 25° is given by

$$50e^{-kt} = 25$$

or $$t = \frac{1}{k}\log 2 = \frac{5\log 2}{\log 1{\cdot}25} = 16{\cdot}0.$$

Hence the time taken for the temperature to fall from 60 to 45 °C is 11 minutes.

We see from equation (ii) that the excess temperature decays exponentially to zero; that is, the actual temperature approaches, but never quite reaches the temperature of the room. This is shown in Figure 6.

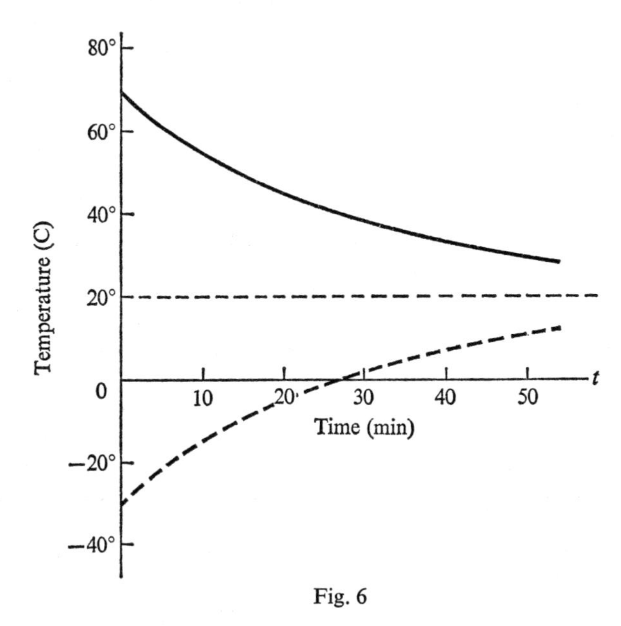

Fig. 6

A similar situation would have existed if the object had originally been at a temperature below that of its surroundings. Equation (i) would still be true, although with $T(t)$ negative, $T'(t)$ positive. A would now be negative and the graph would show an asymptotic approach to room temperature rising from below in the manner shown by the broken curve in Figure 6.

Exercise D

1. It is observed that a population of moths is growing at a rate equal to 0·3 times their present numbers, in moths per week. By what is the population multiplied in 3 weeks? How long would it take for the population to increase by $\frac{1}{10}$th? [$e^{0\cdot 3} = 1\cdot 350$.]

2. A hospital patient whose temperature is 38 °C is in a ward at temperature 21 °C. How long should the thermometer be left in his mouth in order to find his temperature within 0·1° C, if the constant of proportionality governing Newton's law of heating in this case is 5·1, when time is measured in minutes? (Let 38 °C be the zero of the temperature scale you use, so that the room's temperature is −17 °C.) [$e^{1\cdot 7} = 5\cdot 474$.]

3. A group of shipwrecked mariners, 100 strong, are stranded on a desert island with nothing to eat but themselves and take to cannibalism. Taking the rate of consumption of cannibal to be proportional to the amount of cannibal there, and regarding cannibal as a continuous medium, how long, in terms of the constant of proportionality, is it before there is only one cannibal left? [$\log 100 = 4{\cdot}605$.]

4. X g of a chemical A are mixed with a large quantity of chemical B, under whose influence it turns into chemical C. The rate at which it changes in g/s is directly proportional to the amount of A remaining. Find how long, in terms of the constant of proportionality, it takes to reduce the the quantity of A to λX g. Sketch the graph of this time as a function of λ.

5. A heated pellet of iron is plunged in a large bath of water at 20 °C. If the iron is originally at 70 °C and it takes 4 minutes to fall to 25 °C, what is the rate of fall of temperature per unit difference in temperature? [$\log 10 = 2{\cdot}3026$.]

6. A cylindrical drum of oil leaks away through a small hole in the bottom at a rate proportional to the fluid pressure at the bottom, that is, to the depth of oil still in the drum. Find how long it takes to reduce the amount of oil in the drum to $\frac{1}{10}$th of what it was originally, in terms of the constant of proportionality; and prove that it takes three times as long altogether to reduce it to $\frac{1}{1000}$th.

7. Gas diffuses itself through the fabric of a balloon at a rate proportional to the difference in pressure inside and outside the balloon. If the volume of the balloon remains constant, and it is enclosed in a comparatively large vacuum chamber, how long is it, in terms of suitably chosen constants, before the mass of gas within the balloon is reduced by $\frac{1}{10}$? (Boyle's law states that $pv = km$, where p is the pressure, v the volume, m the mass and k a constant determining the nature of the gas.)

8. A condensor of capacity C discharges a charge q through a resistance R. What is the half-life of the charge on the condensor?

9

TRIGONOMETRICAL FUNCTIONS

1. SINE AND COSINE FUNCTIONS

The reader will already have come across the definition of a sine. In its geometrical context the idea is a familiar one. However, it is not so obvious that *sine* and *cosine* have an importance quite beyond geometry, that in calculus they enable us to investigate a completely new function. For, where x is a real number, we have only to *define* f to be the function which maps x onto $\sin x°$, for us to have a mapping of an entirely new character. Note that $\sin x°$ is a ratio, that is a pure number, so that the mapping is of one set of numbers onto another. The function which we have defined then is a perfectly proper one.

1.1 Graph of $\sin x°$. The graph of the function we have defined, $f(x) = \sin x°$, is the ordinary sinusoidal curve shown in Figure 1. And it is the character of this graph, as much as the geometrical significance of $\sin x°$, which gives the sine function its peculiar importance in mathematics. We have a wave-form lying between 1 and -1. In any two successive intervals of length 360 the pattern of the graph will repeat itself. (Note that this is true even if the intervals do not begin at multiples of 90.)

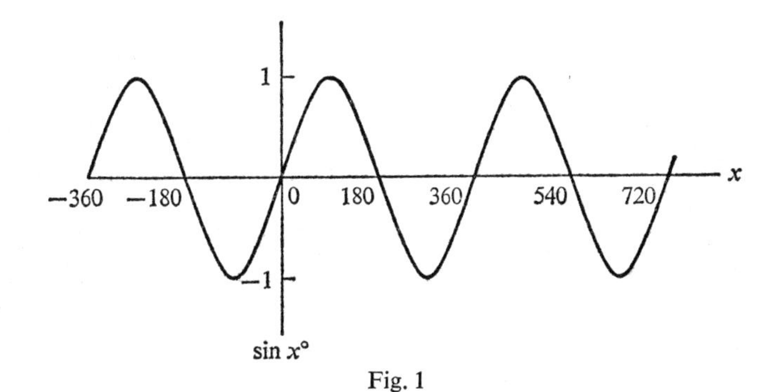

Fig. 1

Functions of this kind we call *periodic functions*, and the interval after which they repeat themselves we call the *period*. Thus in general, if F is a function such that $F(x+k) = F(x)$ for all x, and there is no constant less than k for which it has this property, we say F is periodic with period k. In the present case $f(x+360) = \sin (x+360)° = \sin x° = f(x)$, for all x.

Hence the period is 360. The sine and cosine functions are the simplest of all periodic functions which are given by a single formula. It is this that gives them their special importance in physical work where oscillating functions—as in waves— are of frequent natural occurrence.

1.2 Derivative of sin $x°$. An examination of the form of the graph of $f(x) = \sin x°$ will lead us quickly to a rough idea of its derivative. In the first place, we may note that the tangent at the origin is of small gradient: when x increases by 90, $f(x)$ only increases by 1; and indeed an accurate scale drawing will show us that the gradient, $f'(0)$, is approximately 0·0175. After this, as x increases to 90, as we can see from the original graph of f in Figure 1, $f'(x)$ decreases to 0. As x goes to 180, it further decreases to $-0·0175$: numerically equal to the original value from the symmetry of the curve about $x = 90$. The process is then reversed, and as x moves from 180 to 360, $f'(x)$ returns, through 0, to the original value 0·0175.

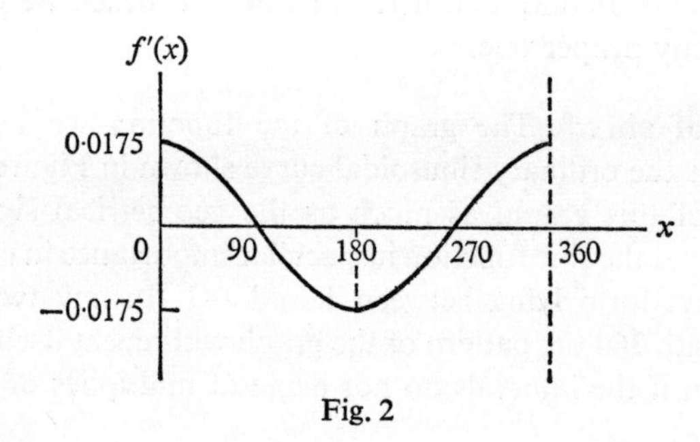

Fig. 2

Figure 2 shows what the graph of $f'(x)$ would look like. After $x = 360$, the cycle repeats itself; and we are lead to conclude, from the whole shape of the graph of f', that it behaves very much as the cosine function, except that whereas $\cos x°$ moves from 1 to -1 to 1 again in the cycle, $f'(x)$ moves between the values $\pm 0·0175$.

We might guess then that $f'(x)$ is $k \cos x°$, where k is a constant to which 0·0175 is an approximation, and whose exact value is $f'(0)$. This is indeed so. We can verify the guess by methods akin to those used in Chapter 7, Section 1.2. For if we define k to be $f'(0)$ (assuming that it exists), then by the definition of derivative,

$$\sin \epsilon° = k\epsilon + s_\epsilon;$$

further it is clear from the graph of $g(x) = \cos x°$ that $g'(0) = 0$ (assuming $g'(0)$ exists). Hence, by the linear approximation near this point,

$$\cos \epsilon° = 1 + s_\epsilon,$$

since $\cos 0° = 1$.

Hence
$$\begin{aligned} f(x+\epsilon) &= \sin (x+\epsilon)^\circ \\ &= \sin x^\circ \cos \epsilon^\circ + \cos x^\circ \sin \epsilon^\circ \\ &= (1+s_\epsilon) \sin x^\circ + (k\epsilon + s_\epsilon) \cos x^\circ \\ &= \sin x^\circ + k \cos x^\circ \epsilon + s_\epsilon \\ &= f(x) + k \cos x^\circ \epsilon + s_\epsilon; \end{aligned}$$
whence
$$f'(x) = k \cos x^\circ.$$

We could also prove this by the method of functional relationship. If $f(x) = \sin x^\circ$, and $g(x) = \cos x^\circ$,

$$f(x+3) = f(x)\, g(3) + g(x) f(3).$$

Assuming $f(x)$ and $g(x)$ to be differentiable, it follows that

$$f'(x+3) = f'(x)\, g(3) + g'(x) f(3).$$

Now $f'(0) = k$, $g'(0) = 0$. Hence, putting $x = 0$. $f'(3) = kg(3)$. In the same way it could be proved for any a that $f'(a) = kg(a)$, which is the same result as before.

2. STANDARD TRIGONOMETRICAL FUNCTIONS

2.1 Radian measures. So far we have defined a function using fundamentally geometrical ideas; and at the basis of these lie the Babylonian measure of angle, leading us to a constant k. Now the constant k plays much the same part as its equivalent with which we dealt when finding the derivative of 2^x; and in just the same way we can dispense with it. For in exponential functions the size of k depended on the arbitrary choice of 2 as the base of our function; here it depends on the fact that our unit of angle is $\frac{1}{360}$th part of a whole rotation. There was nothing essential about the number 2. Equally now there is nothing essential about 360. A more natural unit of angle would be the complete revolution itself; and if we defined a function

$$s(x) = \sin 360x^\circ$$

we would in fact be using this unit. The graph of s would be one in which an increase of one unit in x would produce a complete oscillation (see Figure 3). And the derivative would be,

$$s'(x) = 360k \cos 360^\circ \approx 6{\cdot}3 \cos 360x^\circ,$$

so that the initial gradient would be approximately 6·3.

However in mathematics the apparently natural unit is not always that which we use. Our real requirement in such cases, as we found when

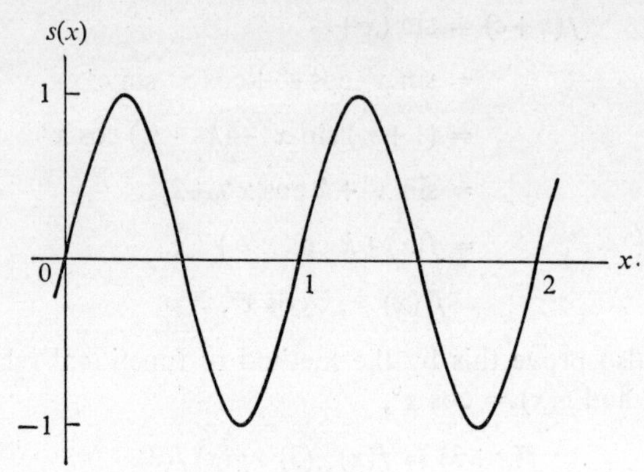

Fig. 3

dealing with exponential functions, is that constants such as k emerging in the course of differentiation, should if possible be made equal to 1; and we can obtain this by choosing a unit of angle, called a *radian*, equal to $(1/k)^\circ$. For if we define the function S by the equation.

$$S(x) = \sin (x \text{ radians}) = \sin (x/k)^\circ,$$

we can see that,

$$S'(x) = k \cos (x/k)^\circ \frac{1}{k} = \cos (x \text{ radians}).$$

If we define C by the equation $C(x) = \cos (x \text{ radians})$,

$$S' = C,$$

and, as we can prove by a method similar to that of Section 1.2,

$$C' = -S.$$

These are the standard trigonometrical functions; and radians are, in consequence, our essential measure of angle. We shall see in the next section a little more fully the relation between radians and degrees. For the moment we can observe that 1 radian $= (1/k)^\circ \approx (1/0{\cdot}0175)^\circ = 57^\circ$ to the nearest degree. (In fact there are somewhat over six radians to a complete revolution.) However it is worth asserting here the importance of radians, an importance which derives from the significance of trigonometrical among the other functions. In future, when we write $\sin x$, of any other trigonometrical ratio of x, we shall assume that x is measured in radians, so that we can now properly write

$$S(x) = \sin x, \quad C(x) = \cos x,$$

and use the symbols sin and cos to denote without ambiguity our standard trigonometrical functions.

2.2 Radians and degrees. So far we have defined 1 radian to be $(1/k)^\circ$ or approximately 57°. However we can establish accurately the relation between radians and degrees by considering the length of the circumference of a circle.

The perimeter of a regular polygon inscribed in a circle gives an approximation to the circumference which can be made as good as we like by increasing the number of sides sufficiently.

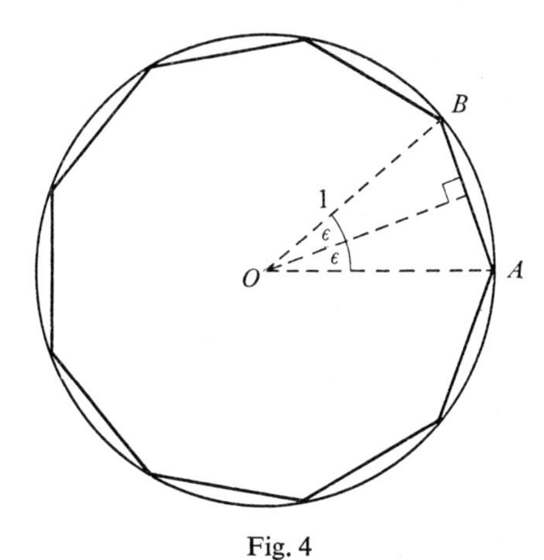

Fig. 4

For convenience we shall take the radius to be 1 unit. See Figure 4. If AB is one of the sides and it subtends an angle of 2ϵ radians at the centre, then $AB = 2\sin\epsilon$.

Now suppose $360^\circ = P$ radians. It follows that there will be $P/2\epsilon$ sides and that the polygon's perimeter will be

$$\frac{P}{2\epsilon}\,2\sin\epsilon = P\frac{\sin\epsilon}{\epsilon}.$$

But $\sin\epsilon = \epsilon + s_\epsilon$, as we have already seen. Hence

$$\frac{\sin\epsilon}{\epsilon} = 1+\frac{s_\epsilon}{\epsilon}.$$

We can make s_ϵ/ϵ as small as we like, and therefore $(\sin\epsilon)/\epsilon$ as near to 1 as we like, by making ϵ small enough. This implies that the perimeter of the polygon can be made as near to P as we like by giving it a large enough number of sides. From this we may reasonably infer that the circumference of the circle is P.

But from the definition of π as the ratio of the circumference of a circle to its diameter, the circumference of this circle is 2π, and therefore $P = 2\pi$: that is, 2π radians $= 360^\circ$ ($2\pi = 6{\cdot}283$ to 3 decimal places).

This completes our idea of radian measure. A simple table of equivalence is:

$$X^\circ = x \textit{ radians}$$

X:	0	30	45	60	90	120	135	150	180
x:	0	$\frac{1}{6}\pi$	$\frac{1}{4}\pi$	$\frac{1}{3}\pi$	$\frac{1}{2}\pi$	$\frac{2}{3}\pi$	$\frac{3}{4}\pi$	$\frac{5}{6}\pi$	π

X:	210	225	240	270	300	315	330	360
x:	$\frac{7}{6}\pi$	$\frac{5}{4}\pi$	$\frac{4}{3}\pi$	$\frac{3}{2}\pi$	$\frac{5}{3}\pi$	$\frac{7}{4}\pi$	$\frac{11}{6}\pi$	2π

We can sketch the graphs of $C(x)$ and $S(x)$ as below in Figure 5, noting the relation between their gradients, and that the greatest value of $|S'(x)|$ and of $|C'(x)|$ is 1.

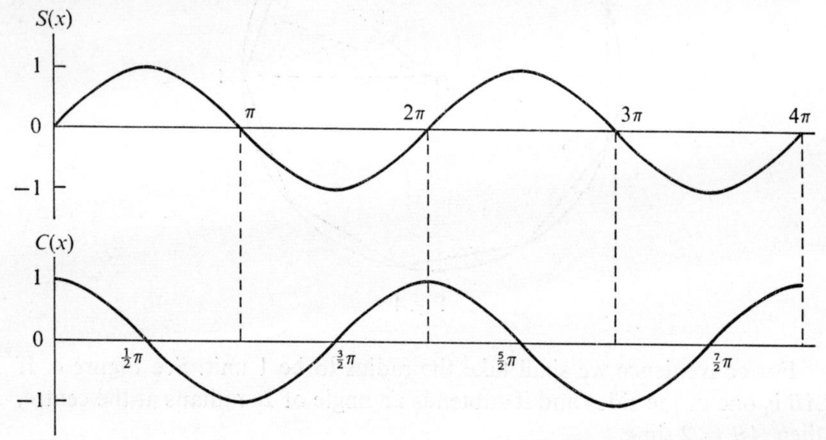

Fig. 5

2.3 Arc-length and area of sector. Let H, K be points on a circle, centre O, and radius r; and suppose angle HOK is θ radians.

The complete circumference subtends an angle of 2π radians at the centre and is of length $2\pi r$; so in proportion,

arc-length HK = rθ.

[The name *radian* arose as the angle at the centre of a circle which is subtended by an arc-length equal to the *radius*.]

The area of a sector of a circle also has a simple formula when the angle

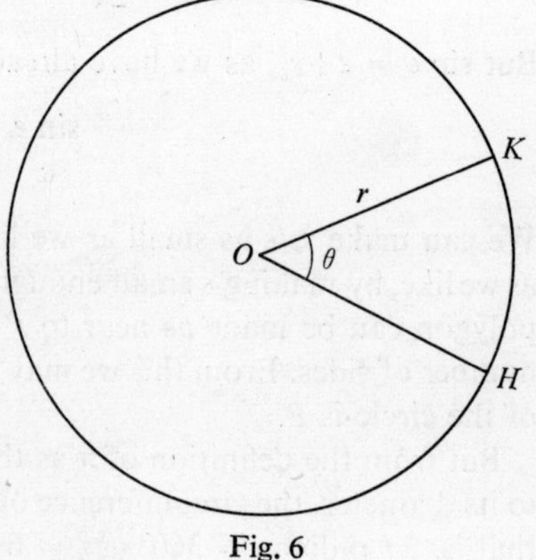

Fig. 6

is expressed in radian measure. The area of the circle (πr^2) is the area of a sector which has an angle of 2π radians at the centre; so, in proportion, the area of the sector whose angle is θ radians at the centre is $(\theta/2\pi)\,\pi r^2$.

Hence the **area of the sector HOK** $= \frac{1}{2}r^2\theta$.

2.4 Simple trigonometrical functions. Sin x, as we have said before, has an importance in mathematics quite apart from its geometrical significance, as a particularly simple periodic function. Its period is 2π; for if x changes by this value sin x will remain unchanged. However we can use

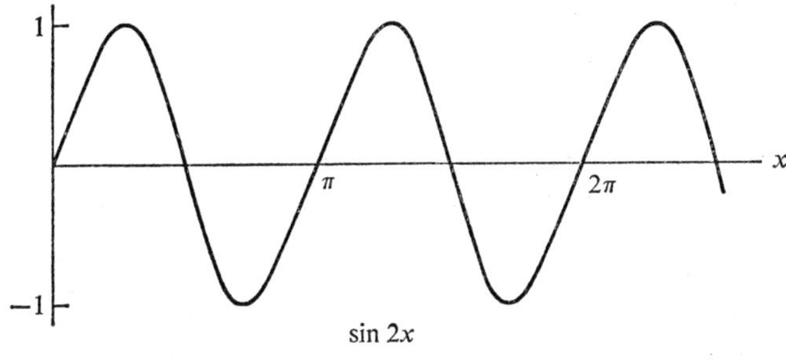

sin $2x$

Fig. 7

this to obtain a similar periodic function of any other period. Sin $2x$ for instance has a period π; for if x moves from 0 to π, $2x$ moves from 0 to 2π, and sin $2x$ performs a complete oscillation. Its graph is shown in Figure 7. Likewise sin $x/3$ has a period 6π, since it requires this change in x for $x/3$ to change by 2π; and in general sin kx has a period $2\pi/k$.

If
$$f(x) = \sin 2x,$$
$$f'(x) = \cos 2x.2.$$

Consequently its gradient where $x = 0$ is 2: twice that of sin x. We can think of the graph of sin $2x$ in fact as that of sin x contracted by a factor 2 in the horizontal direction.

If $f(x) = \sin(x+3)$, we have a different kind of transformation of the graph. Here the period is still 2π, since when x changes through this value, so also does $x+3$. But as x moves from 0 to 2π, $x+3$ moves from 3 to $2\pi+3$, so that we have a complete oscillation starting from a point other than the neutral point. We have in fact the graph of sin x translated a distance 3 units in the direction shown. See Figure 8. (The reader will think of 3 radians as about 170°, or a little less than half a complete period.)

This suggests to us relations between the sine and cosine functions. Evidently the graph of $C(x)$ is that of $S(x)$ translated a distance $\frac{1}{2}\pi$ units to the left. Hence,
$$C(x) = S(x+\tfrac{1}{2}\pi),$$
or
$$\cos(x) = \sin(x+\tfrac{1}{2}\pi),$$

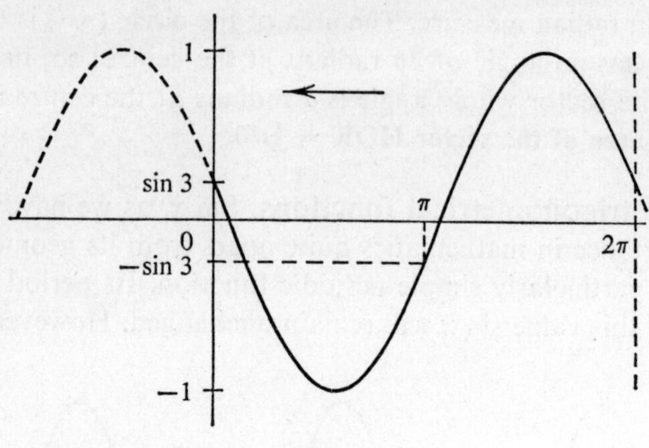

Fig. 8

a geometrical result, which, expressed in degree measure, the reader may well have met before. This shows that the sine and cosine functions are essentially the same as far as the character of their graphs go, and as we have already said it is this character that establishes a great deal of their importance in mathematics. It is useful for manipulative purposes to be familiar with both, but in practice we can often think of them as representing the same oscillating function, in forms however which differ by $\frac{1}{2}\pi$ in their phase.

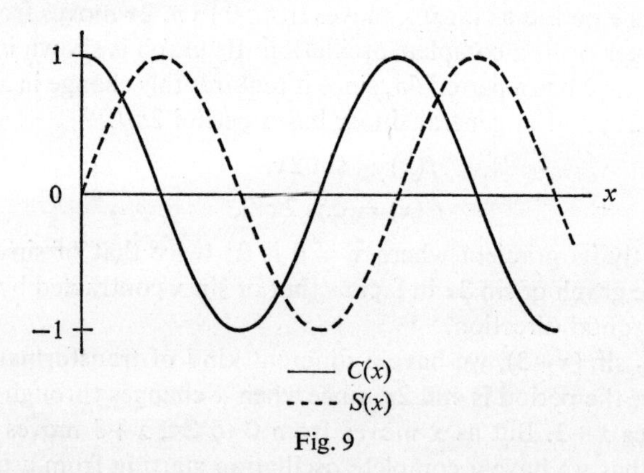

Fig. 9

Example. Sketch the graph of the function $5 \sin (4x-\pi)$.

We can think here first of the graph of $\sin 4x$, which is that of $\sin x$ contracted by a factor 4; then $\sin (4x-\pi) = \sin 4(x-\frac{1}{4}\pi)$, which is that $\sin 4x$ translated a distance $\frac{1}{4}\pi$ units to the right; and lastly $\sin (4x-\pi)$ multiplied by a factor 5.

We can verify this by observing independently that $5 \sin (4x-\pi)$ is an oscillating function with period $\frac{1}{2}\pi$, and that it is zero when $x = \frac{1}{4}\pi$.

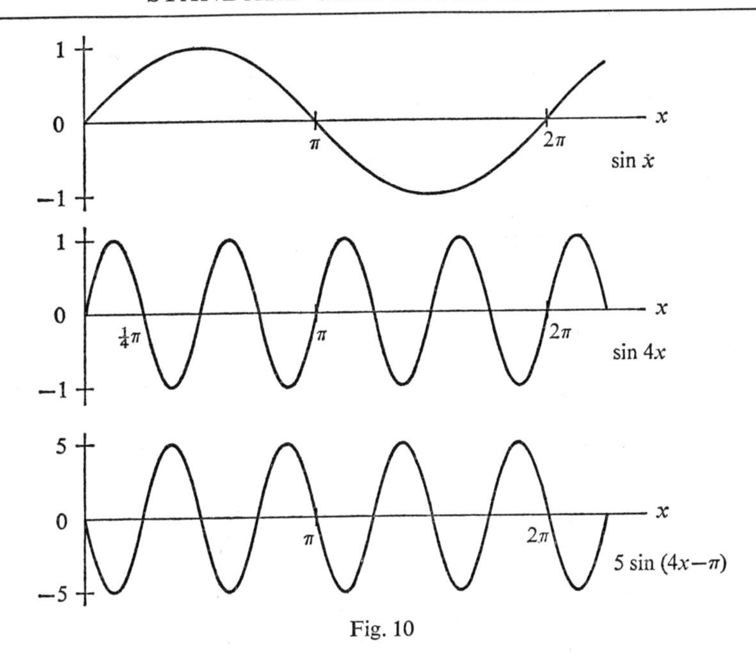

Fig. 10

Exercise A

1. Convert to radians (in terms of π): 180°, 90°, 270°, 60°, 120°, 30°, 150°, 330°.

2. Convert to degrees: $\frac{1}{4}\pi$, $\frac{2}{3}\pi$, $\frac{1}{6}\pi$, $\frac{3}{4}\pi$, $\frac{4}{5}\pi$ radians.

3. For a circle of radius 10 cm, find (i) the arc-lengths which subtend at the centre angles of: 1, 2, $\frac{1}{3}\pi$ radians; (ii) the areas of the sectors whose angles at the centre are: 1, $\frac{1}{2}$, $\frac{2}{5}\pi$ radians.

4. What is the ratio of the lengths of the chord *HK* to the arc *HK* of a circle where *HK* subtends an angle at the centre of (i) $\frac{1}{10}\pi$; (ii) $\frac{1}{100}\pi$?

5. Sketch the graphs of the following functions:

$$\sin 3x^\circ,\quad \sin(x+40)^\circ,\quad \cos(2x-120)^\circ,\quad \sin(-x)^\circ,$$
$$\cos(-x)^\circ,\quad \sin(270+3x)^\circ,\quad \cos(40-\tfrac{1}{2}x)^\circ.$$

Estimate their gradients where they cross the x-axis.

6. Differentiate the following functions and sketch their graphs:

$$\sin 3x,\quad \cos\tfrac{1}{4}x,\quad \cos 2x,\quad 3\sin\tfrac{1}{2}x,\quad \sin(x+\pi),\quad \cos(2x-\tfrac{1}{2}\pi),$$
$$5\sin(\tfrac{2}{3}\pi-4x),\quad 3(1-\cos 3x),\quad 3\sin(\tfrac{3}{4}x+\tfrac{1}{2}\pi)-2.$$

7. Prove (i) by the method of linear approximation, and (ii) by the method of function relationship (see Section 1.2) that $\cos' = -\sin$.

8. Use linear approximation to get a value for cos 59°. (Note that it is best to put this angle into radians.)

9. Use Newton's method to obtain the root of $\sin x = 0{\cdot}4$ near $\frac{1}{6}\pi$ correct to two decimal places. Find a root also for $\cos x = -0{\cdot}4$ to two decimal places.

3. FURTHER TRIGONOMETRICAL FUNCTIONS

Hardly less important than the primary sine and cosine functions are those which directly depend on them: the secant and tangent functions, with cosecant and cotangent, and many which can be expressed partly in terms of them. In this section we shall discuss some of these, and the behaviour of their graphs.

3.1 Sec x and its derivative. Suppose $f(x) = \sec x$. Knowing the graph of $\cos x$ it is easy to sketch that of the secant function, since $\sec x = 1/\cos x$, and it is sufficient for us to consider the reciprocals of values shown in the first. Figure 11 shows the relationship.

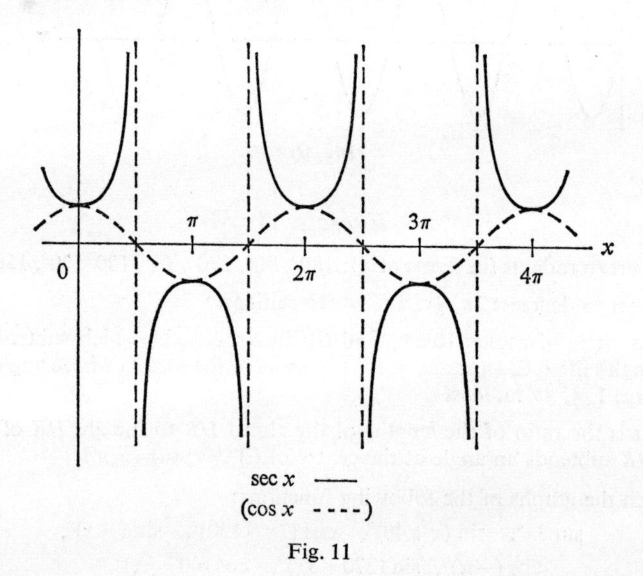

Fig. 11

The derivative of sec x can be found in a like manner. Regarding $f(x)$ as $P_{-1}(\cos x)$, we see that

$$\begin{aligned} f'(x) &= P'_{-1}(\cos x)\,(-\sin x) \\ &= -(\cos x)^{-2} \times (-\sin x) \\ &= \frac{1}{\cos x} \times \frac{\sin x}{\cos x} \\ &= \sec x \tan x. \end{aligned}$$

The graph of this can readily be inferred from that of sec x. It is shown in Figure 12, and the reader is left to check the correspondence for himself. It could also be obtained by 'multiplying' the graphs of tan x and sec x; but for that we must turn to the next section.

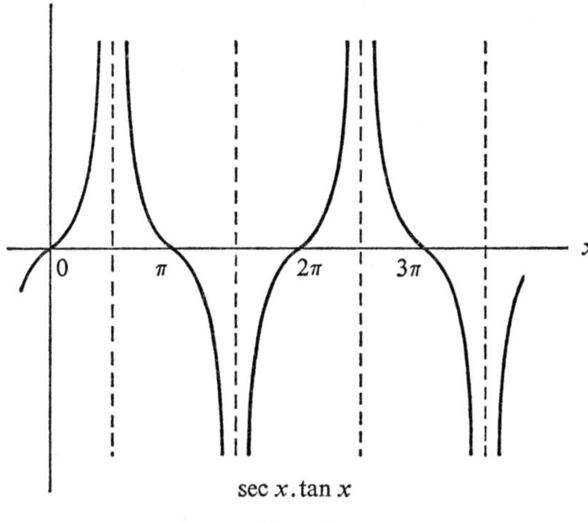

sec x.tan x

Fig. 12

3.2 Tan x and its derivative. The graph of tan x likewise can be found from those of more elementary functions.

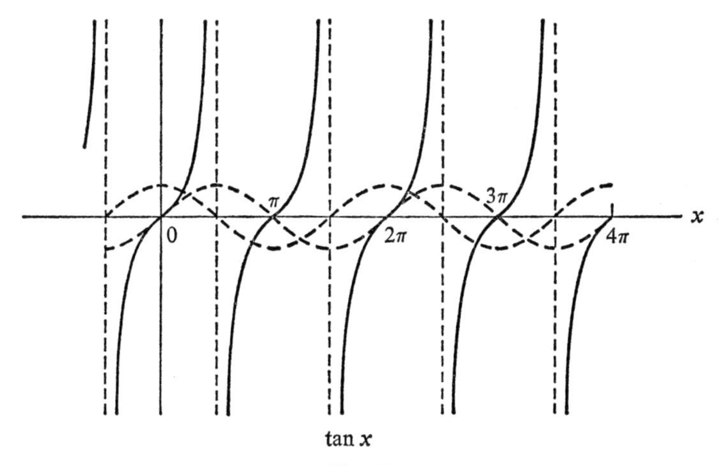

tan x

Fig. 13

We know that

$$\tan x = \frac{\sin x}{\cos x}$$

for values of x not equal to odd multiples of $\frac{1}{2}\pi$. If we sketch the graphs of these two and consider their ratios, it is clear that $\tan x$ must take the form shown, tending to $\pm\infty$, whenever x approaches an odd multiple of $\frac{1}{2}\pi$ (since here $\cos x = 0$), and periodic with period π. It is clear also that the graph has point symmetry about 0, since $\tan(-x) = -\tan x$; hence the inflection at 0.

For its derivative we observe that,

$$f(x) = \tan x = \frac{\sin x}{\cos x} = \sin x \sec x;$$

hence

$$\begin{aligned} f'(x) &= \cos x \sec x + \sin x \sec x \tan x \\ &= 1 + \tan^2 x, \qquad \text{since } \sin x \sec x = \tan x, \\ &= \sec^2 x. \end{aligned}$$

The graph of this may be inferred from that of $\sec x$. The fact that it is always positive corresponds, as we would expect, to the continual upwards slope of the curve of $\tan x$.

We may note also at this stage the cotangent and cosecant functions. These are very similar to the tangent and secant, and it can be readily shown that their derivatives are $-\mathrm{cosec}^2 x$ and $-\mathrm{cosec}\, x \cot x$. The proof of these formulae, and the sketching of their graphs, are left as exercises to the reader.

3.3 Composite functions. There are of course a host of simple functions which directly depend on those we have discussed above. We give one or two examples of how they can behave, and the kinds of techniques that are necessary to analyse their behaviour.

Example 1. If $f(x) = \sin^4 x$ find $f'(x)$ and sketch the graph of f.

We can regard $f(x)$ as $P_4(\sin x)$. Then

$$\begin{aligned} f'(x) &= P_4'(\sin x) \cos x \\ &= 4 \sin^3 x \cos x. \end{aligned}$$

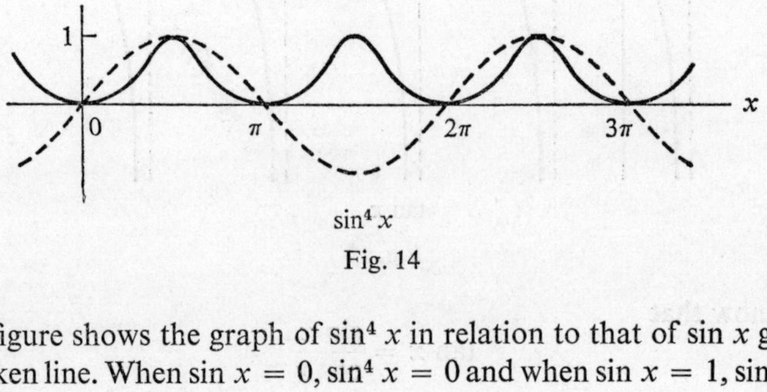

1
0
π
2π
3π
x

$\sin^4 x$

Fig. 14

The figure shows the graph of $\sin^4 x$ in relation to that of $\sin x$ given by the broken line. When $\sin x = 0$, $\sin^4 x = 0$ and when $\sin x = 1$, $\sin^4 x = 1$

also; but when $\sin x = \frac{1}{2}$, for example, $\sin^4 x = \frac{1}{16}$—a much smaller value—and we therefore expect the graph of $\sin^4 x$ to have a set of very flat minimum points where x is zero or any integral multiple of π. The turning points occur where $f'(x) = 0$: that is where $\sin x = 0$ or $\cos x = 0$. At these $f(x)$ has its extreme values of 0 or 1.

Example 2. Sketch the graph of $e^{\tan x}$ and discuss its behaviour near $x = \frac{1}{2}\pi$.

If $f(x) = e^{\tan x}$ then $f'(x) = e^{\tan x}\sec^2 x$. For values of x greater than but near $\frac{1}{2}\pi$, $e^{\tan x}$ is small and $\sec^2 x$ is large. As $x \to \frac{1}{2}\pi$, if $x > \frac{1}{2}\pi$, $\tan x \to -\infty$ and therefore $e^{\tan x}\tan^2 x \to 0$ (see Chapter 7, Section 3.3). But $e^{\tan x}\sec^2 x = e^{\tan x}(1+\tan^2 x)$. Hence $f'(x) \to 0$.

As $x \to \frac{1}{2}\pi$, $f(x) \to 0$ if $x > \frac{1}{2}\pi$, but $f(x) \to \infty$ if $x < \frac{1}{2}\pi$. Hence the graph in Figure 15.

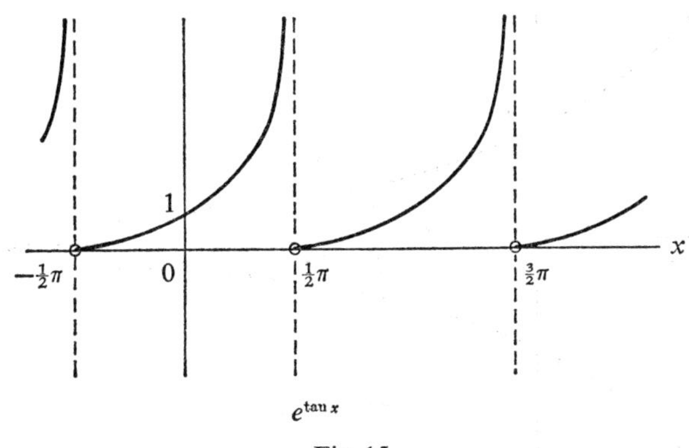

$e^{\tan x}$

Fig. 15

Example 3. Sketch the graph of $\log \sin x$. For what values of the domain is it an increasing function?

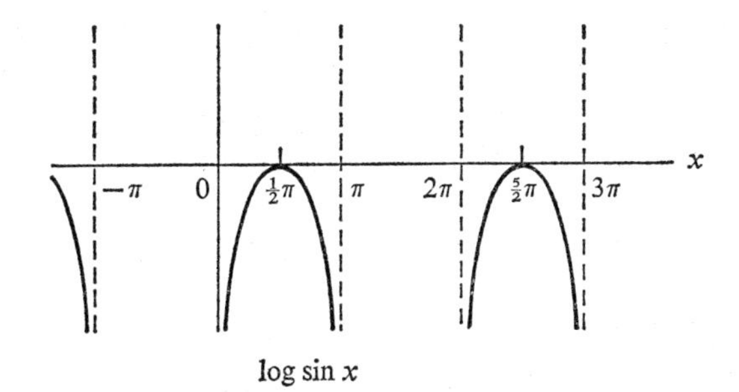

$\log \sin x$

Fig. 16

If
$$f(x) = \log \sin x$$

then
$$f'(x) = \frac{1}{\sin x} \cos x = \cot x.$$

The logarithm of a negative number is undefined so that $f(x)$ (and therefore $f'(x)$) is undefined in the intervals $(2n-1)\pi$ to $2n\pi$, where n is any integer. $\cot x \geqslant 0$ in the intervals $2n\pi$ to $(2n+\frac{1}{2})\pi$, so that $f'(x) \geqslant 0$ there. Hence f is an increasing function in these intervals; but not of course in the other intervals where $\cot x$ is positive, as f is not defined there.

Example 4. Sketch the graph of $f(x) = e^{-x/2\pi} \cos x$.

$|\cos x| \leqslant 1$, so that $f(x)$ lies always between $e^{-x/2\pi}$ and $-e^{-x/2\pi}$. The graphs of these last two functions of x are shown in dotted lines in

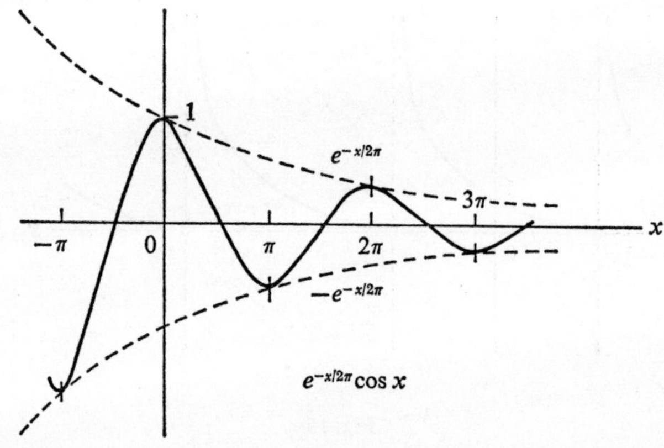

Fig. 17

Figure 17. $e^{-x/2\pi}$ is never zero, so that $f(x)$ is zero only where $\cos x$ is zero; that is, where x is an odd multiple of $\frac{1}{2}\pi$.

If x is an integral multiple of π, $\cos x$ is 1 or -1, so that here $f(x) = \pm e^{-x/2\pi}$. It follows that, since the graph of $f(x)$ is bounded by those of $e^{-x/2\pi}$ and $-e^{-x/2\pi}$, it must touch one or other of these graphs at these points.

This can be verified from the derived function:

$$f'(x) = -e^{-x/2\pi}\left(\frac{1}{2\pi}\cos x + \sin x\right).$$

This is equal to $-(e^{-x/2\pi}/2\pi)$ when x is an even multiple of π, and $(e^{-x/2\pi}/2\pi)$ when x is an odd multiple of π. These are the same as the derivatives of $e^{-x/2\pi}$ and $-e^{-x/2\pi}$ respectively.

These points obviously do not give turning values of f since the gradients are not zero. The turning points occur where $f'(x) = 0$, that is where $\tan x = -1/2\pi$. These occur then where $x = -\alpha + n\pi$, where

$$\alpha = \tan^{-1}(1/2\pi).$$

The graph is shown in Figure 17. Functions of this kind arise in connection with lightly damped oscillations.

Example 5. Sketch the graph of $f(x) = x \sin x$.

Figure 18 shows the graphs of x and $\sin x$. Where x is small, $\sin x \approx x$ and therefore $x \sin x \approx x^2$, and this gives us the shape of the graph of f near the origin.

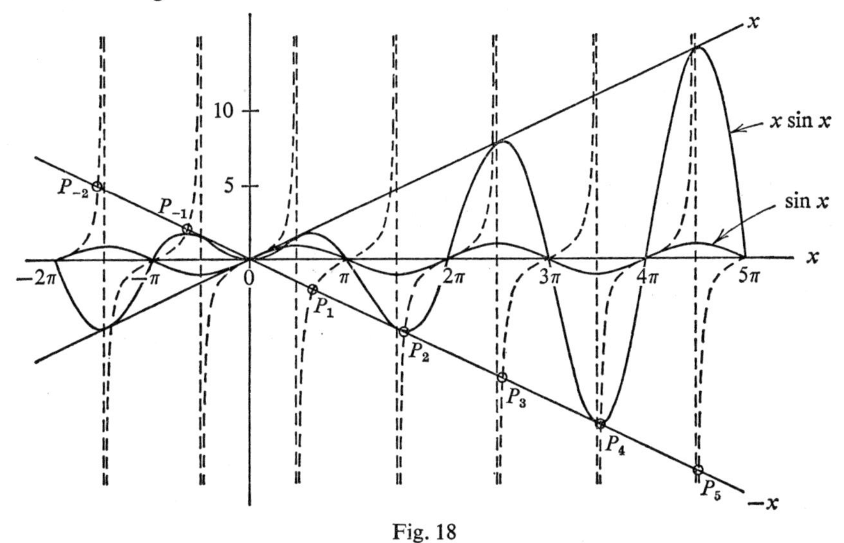

Fig. 18

Where $x = \frac{1}{2}\pi$, $x \sin x = \frac{1}{2}\pi$; and near $x = \frac{1}{2}\pi$, on either side, $\sin x < 1$, so that the graph of $x \sin x$ will touch the graph of x at this point. We can confirm this by observing that

$$f'(x) = \sin x + x \cos x$$

and $f'(\frac{1}{2}\pi) = 1$.

In the same way the graph of f will touch either the graph of x or the graph of $-x$ (also shown in the figure) wherever x is an odd multiple of $\frac{1}{2}\pi$; so clearly the turning points of $x \sin x$ do not correspond to the turning points of $\sin x$. In fact, f has its turning points where $f'(x) = 0$; that is, where $\tan x = -x$.

Figure 18 also shows the graph of $\tan x$ (as a broken line), and the points where this meets the graph of $-x$ are marked $P_1, P_2, P_3, \ldots$. The turning points of f lie on the vertical lines through these points. We also have the conditions that the zeros of f occur where $\sin x = 0$: that is, at multiples of π; and now we are in a position to make a sketch of f.

Note: the graph of f exhibits the characteristics of *forced oscillations*. For instance, if the rim of a wine-glass is tapped, it rapidly vibrates or oscillates with a frequency depending on the size, shape and material properties of the glass, so that it emits a sound wave of that frequency. If a singer, near the glass, pitches his voice to produce a note of exactly the same frequency, this also sets up oscillations in the glass; and since these are caused by an external force oscillating with the same frequency as the *natural* frequency of the glass, their amplitude steadily increases as shown in Figure 18. This causes the note to grow louder and can, eventually, crack the glass. The phenomenon—which can occur in other physical systems as well—is known as resonance. It has useful applications in electric circuits.

Exercise B

1. What are the derivatives of $\operatorname{cosec} x$, $\sin^2 x$, $\tan 2x$, $\sec^2 x$, $\cos^3 2x$, $\cot x$, $\sin x \cos x$, $x^2 \tan 3x$, $\log \sin x$, $\sin (e^x)$, $\log \sec 2x$?

2. If $f(x) = \sin^2 x$ and $g(x) = \cos^2 x$, what are $f'(x)$ and $g'(x)$, and what can you deduce from this?

3. Sketch the graphs of (i) $\tan 3x$; (ii) $\sec^2 x$; (iii) $\cos^3 x$; (iv) $\sec (\frac{1}{4}\pi + x)$.

4. Sketch the graphs of (i) $x \cos x$; (ii) $x^2 \cos 2x$; (iii) $(\sin x)/x$.

5. Sketch the graphs of (i) $e^{-\frac{1}{2}x} \sin x$; (ii) $e^{3x} \cos 4x$.

6. Sketch the graphs of (1) $e^{\sec x}$; (ii) $\sin (e^x)$; (iii) $\log \sec x$.

7*. Sketch the graphs of (i) $\sin (1/x)$; (ii) $x \sin (1/x)$.

8. Sketch the graphs of:

$$\cos^2 x, \quad 1+\cos^2 x, \quad (1+\cos^2 x)^{\frac{1}{2}}, \quad (1+\cos^2 x)^{-\frac{1}{2}} \quad \cos 4x(1+\cos^2 x)^{-\frac{1}{2}}.$$

9. Sketch the graphs of $\sin x - 2x$, $\sin x - x$, $2 \sin x - x$. How many roots has the equation $2 \sin x - x = 1$?

10. How many positive roots has the equation $e^x = 9 \sin x$?

11. How many roots has the equation $6 \cos x = x$?

12. What is the area contained between the x-axis and a single hump of the sine curve?

13. The area under the curve $y = \sec x$ from $x = -\frac{1}{4}\pi$ to $x = \frac{1}{4}\pi$ is revolved round the x-axis. What is the volume defined in this way?

14*. How many roots has $\tan x = 2x$ between 0 and 2π? Calling the roots (if any) $\alpha_1, \alpha_2, \ldots$, discuss the number of roots of the equation $\sin^2 x = ax$ between 0 and 2π, for various ranges of values of a.

4. OSCILLATIONS AND WAVES

4.1 Waves. We have already mentioned that a great part of the importance of trigonometrical functions lies in the frequency with which they turn up in nature to represent a *wave* or *oscillation*. In this section we shall consider some of these phenomena. Let us take, to illustrate the fundamental ideas, one of the simplest natural cases: the effects of a pebble dropped into a pond of still water. Here evidently we have a case of *wave motion*; for experience would lead us to expect a series of ripples, travelling outwards in concentric circles, diminishing in height (though not in the intervals between them) both as time passes and as the individual ripples expand, until eventually the pond is smooth.

Now if we take a vertical cross-section of the pond at a particular instant, and over a distance too small to show the lessening of the ripples, we might observe that the surface of the water lay in the form of a sine curve; and this is indeed the mathematically simplest wave formation that can occur.

We call the length of a particular full wave the *wave-length*; and the greatest height of a wave above the mean water level the *amplitude*. Thus if the equation of the curved surface of the water is $y = A \sin px$, the amplitude is A, and, since every time px increases by 2π the curve describes a full wave, the wave-length is $2\pi/p$.

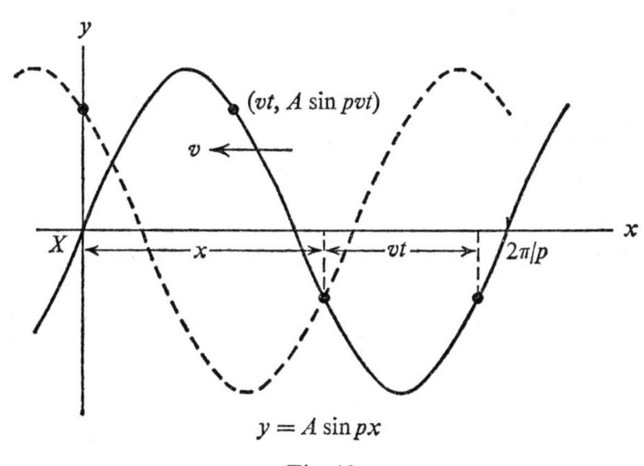

Fig. 19

Consider now what happens to the water level of the pond at a particular point, rather than at a particular instant of time. We have said that the ripples move outwards, and in the simplest wave formation they will move with a constant speed (v say) while at the same time preserving exactly their shape. To see therefore what happens to the water level at a point, we have only to imagine the curve in Figure 19 move past the point X at constant velocity v from right to left. Evidently the water level itself will oscillate. After time t the point on the original curve (vt, $A \sin pvt$) will have moved to a position vertically above X, and hence the height of the water will be

$$y = A \sin pvt = A \sin nt,$$

where n is equal to pv.

We can plot this variation also in a graph, the height now against the *time*, and variations of this kind with respect to time we call *oscillations*.

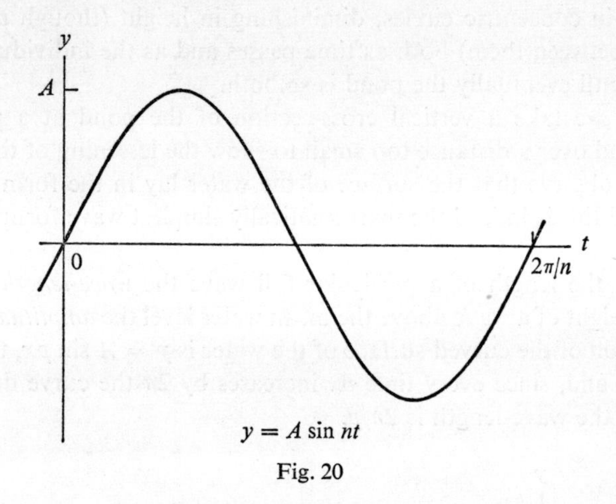

Fig. 20

A is again spoken of as the amplitude of oscillation. The length of time for a complete oscillation we will call the *period*, here $2\pi/n$, or $2\pi/pv$. The number of oscillations per unit time is the *frequency*, here $n/2\pi$. And we see that for a wave motion of this kind,

$$\textit{wave-length} = \textit{velocity} \times \textit{period}, \text{ since } \frac{2\pi}{p} = v \times \frac{2\pi}{pv}.$$

The *general* height of the water is of necessity expressed as a function of two variables, for it depends both on the *position* of the point under consideration, and upon the *time*. If Figure 19 shows the water level when $t = 0$, then after time t the wave will have moved a distance vt; it follows that at the point a distance x from X, the height of water will be what it was at $x+vt$ t units of time before. Hence the height will be

$$y = A \sin p(x+vt)$$

$$= A \sin \frac{2\pi}{\lambda}(x+vt),$$

where λ is the wave-length and v the velocity of the waves.

4.2 Sound. Wave motions of the kind sketched in the last section occur often in the physical world, where there is a natural tendency for quantities to oscillate about their equilibrium position. One of the most important instances is that of *sound waves*. We will discuss these here, both for their intrinsic interest and for the light they throw on all such phenomena.

Sound is caused originally by some rapidly vibrating object, like a violin string, or the prong of a tuning fork, or the diaphragm of a drum. These vibrations cause small disturbances in the air pressing upon the object; in fact each particle of air can be regarded as oscillating rapidly backwards and forwards in the direction in which the sound is travelling. This particle in its turn agitates those that are next to it, and so the disturbance travels from the original source rather as a succession of sudden impacts can be seen travelling down a line of shunting trucks, until it impinges on the listener's ear.

Now, with a pure musical note—for instance that of a flute rather softly blown—this vibration will take almost exactly the waveform sketched in the preceding section. Let us consider a line along which sound is travelling and the disturbances of the air particles backwards and forwards along that line at a particular instant. We shall get a picture rather like that shown in Figure 21.

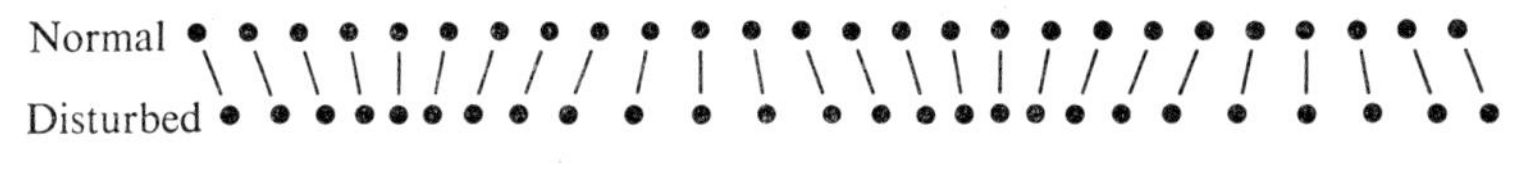

Fig. 21

Where the particles are more crowded the air pressure will be higher, and where they are dispersed it will be lower. Therefore we may think equally of sound waves as waves of variation of air-pressure. To obtain however the best graphical idea of the wave formation, we will plot the

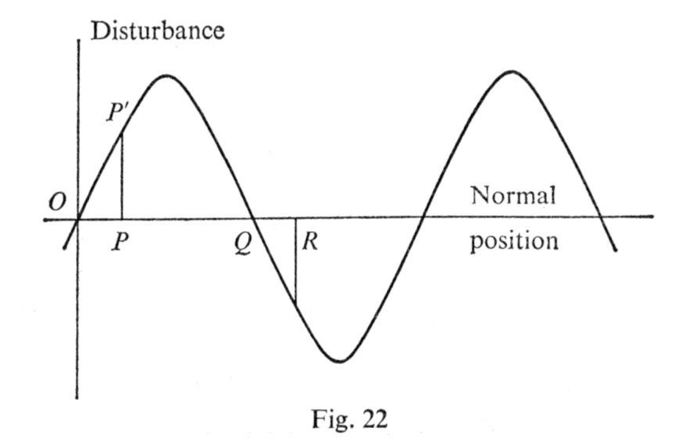

Fig. 22

disturbance of each particle against its normal distance from the source; so that the air particles normally at a distance OP from O, for instance, have been pressed a distance PP' ahead of their normal position; at a distance OQ they are momentarily in their normal position; and those normally at a distance OR have been drawn behind. It may be emphasized that this is a picture of the disturbance at a particular *time*; and in what follows it will be sufficient to consider this instantaneous wave formation; however it is important to remember that it is in fact travelling with the velocity of sound.

The *nature* of the sound will clearly depend on the wave formation. The *pitch* is found to depend on the frequency of oscillation: the higher the frequency the higher the note; and therefore, in the figure we are using, the smaller the wave-length the higher the note. The *loudness* of the note depends also on the frequency, in part; however, for notes of the same pitch, the loudness depends only upon the amplitude of oscillation—obviously the greater the amplitude the greater the disturbance of air, and therefore the greater the effect on the listener's ear.

Sound waves are in fact damped, that is to say their amplitude decreases with distance and with time though their period remains constant. For a simple model of damping, see Section 3.3, Example 4.

4.3 The natural and diatonic scales. Now the *relation* between two notes does not depend only on the pitch at which they sound. For instance, if even a musically naïve reader sings the words 'God save our gracious Queen', he will probably recognize that the notes on 'save' and 'our' are further apart than those on the two syllables of 'gracious'; and that if he goes on to 'Long live our...', the relation between 'live' and 'our' is exactly the same as that between gra-' and '-cious', even although they are sung at a higher pitch.

All this is capable of precise mathematical formulation; for the interval between two notes depends only on the *ratio between their frequencies*. Indeed the ordinary musical notes used for melodic purposes are those whose frequencies are closely related on what is called the 'natural scale'. Consider for instance a simple non-valve trumpet. If the lowest note which a trumpeter can make on it has frequency 1 oscillation per unit time, then by altering the pressure of his lips he can also make the column of air in the instrument oscillate twice as fast, obtaining a frequency 2 oscillations per unit time. Likewise he can obtain notes with frequencies 3, 4 and 5 and so on, by causing the air to oscillate with some constant multiple of the original frequency, which depends only on the size and shape of the trumpet.

These are the notes which are closely related to the original, and which if sounded together with it will produce a harmonious effect. The note with twice the original frequency is that an octave above the original. That with three times as much is a further interval of a fifth above that, and so on. The notes of this natural scale are shown below as they appear on the ordinary keyboard:

Note	C	C′	G′	C″	E″	G″		C‴	*D*‴	E‴
Frequency	1	2	3	4	5	6	7	8	9	10

and these are the only notes which the ordinary non-valve trumpet can make. (The valves on a trumpet are a device for altering the fundamental length of the column of oscillating air, and so obtaining more than one fundamental note.) Note that a frequency 4 again obtains a note an octave above the second C (C′) whose frequency is 2: that is C″. That with frequency 6 is an octave above the lowest G (G′), and so on. That with frequency 7, although the trumpet can make it, makes no appearance in European music and would be merely disconcerting if played. The same, if we went on, would be true of those with frequencies 11, 13 and so on. However the notes at intervals based on frequencies of 2, 3 and 5 are sufficient to provide the basis of the ordinary scale.

For, returning to the scale above the lowest C, G is an octave below G′ which has frequency 3: therefore it has frequency 3/2. Likewise E is two octaves below E″ and has frequency 5/4, and D is three octaves below D‴ and has frequency 9/8. F, A and B are not in the natural scale, but we can get them by protracting the same harmonic system backwards. F stands in the same relation to C′ as C stands to G: that is to say, if we had a natural scale with F as its basic note, C′ would have frequency 3/2 times as great. Hence the frequency of F is 2/3 that of C′, which is 4/3. A stands to E as F stands to C, and it has frequency 5/3. Likewise *B* has frequency 15/8. Hence we have the octave based on the natural scale:

Note	C	D	E	F	G	A	B	C′
Frequency	1	9/8	5/4	4/3	3/2	5/3	15/8	2

If we examine these frequencies, we shall find that adjacent notes of the scale are separated by three different intervals. C to D, F to G and A to B, all are in the ratio 8:9. D to E, G to A, are in the ratio 9:10, which is a smaller interval although so close that it requires a good musical ear to detect the difference. E to F and B to C have ratio 15:16, which is a substantially smaller interval, and can easily be recognized as such. The intervals then of the ordinary scale can be symbolized as below, with three different sizes, two of them nearly the same. It follows that if

$$\mathrm{C} \equiv \mathrm{D} \equiv \mathrm{E}-\mathrm{F} \equiv \mathrm{G} \equiv \mathrm{A} \equiv \mathrm{B}-\mathrm{C}$$

a key-board instrument, a piano say, is correctly tuned for music in the scale of C, it cannot ideally be used for anything else. For instance the scale of D, from D to E, begins with an interval of 9/10 whereas it should begin with 8/9; and to work satisfactorily even in a limited range of keys, there would be needed a whole host of new notes. It was in consequence of this that the system of 'equal temperament', a compromise tuning so that all keys can be played only slightly out of tune, was gradually adopted in the eighteenth century. Only in such media as string quartets and unaccompanied vocal music, where the musicians can rely entirely on their own sense of pitch, can the music be accurately played.

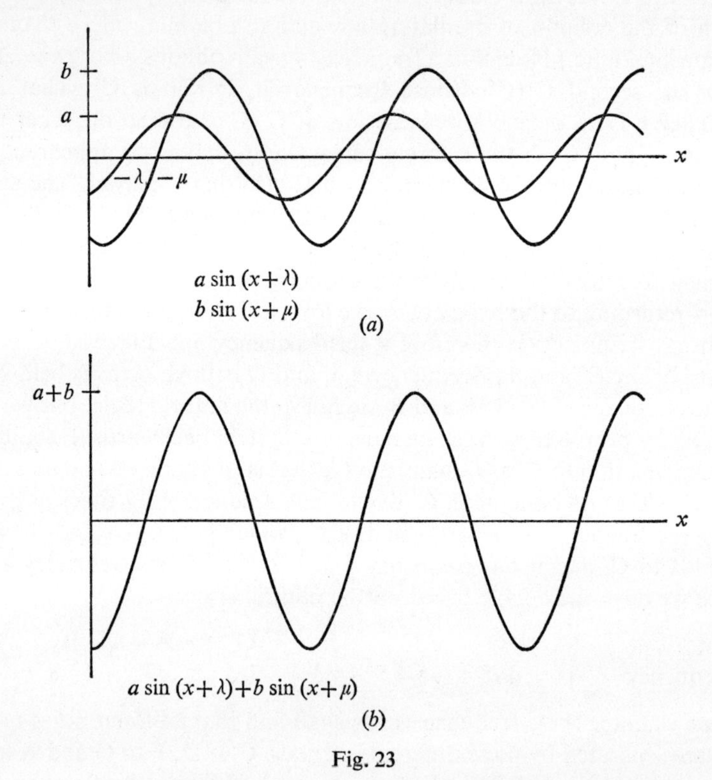

Fig. 23

5. COMPOSITE WAVE FORMATIONS

5.1 Sum of sine functions of the same period. The disturbance in the air caused by two or more pure notes of the same pitch sounding together can be found by adding the individual disturbances. Figure 23*a* shows two such sine waves having the same period but of different amplitudes and of different phases (that is, their zeros do not coincide). Musically one would expect to get from them a single pure note of the same pitch; and indeed the sine functions do add to a sine function of the same period, as shown in Figure 23*b*. Let us consider the sum of functions each of the form $a \sin(x+\lambda)$, where a and λ are constants. This function is periodic with period 2π.

Let
$$f(x) = a \sin(x+\lambda)+b \sin(x+\mu),$$
where a, b, λ, μ, are constants, positive or negative.

If Figure 23*a* represents the graphs of $a \sin(x+\lambda)$ and $b \sin(x+\mu)$, where a and b are the amplitudes and the graphs cross the x-axis where $x = -\lambda$ and $x = -\mu$, as shown, then Figure 23*b* will be the graph of $f(x)$.

We shall show that the sum of two, and therefore of any number of functions like $a \sin(x+\lambda)$ is itself a sine function of the same type; that is
$$a \sin(x+\lambda)+b \sin(x+\mu) = r \sin(x+\alpha).$$
Now,
$$a \sin(x+\lambda) = a \cos\lambda \sin x+a \sin\lambda \cos x$$
$$= p_1 \sin x+q_1 \cos x,$$
where the constants p_1, q_1 are $a \cos\lambda$, $a \sin\lambda$. Similarly
$$b \sin(x+\mu) = p_2 \sin x+q_2 \cos x.$$
Therefore,
$$a \sin(x+\lambda)+b \sin(x+\mu) = p \sin x+q \cos x,$$
where
$$p = p_1+p_2; \quad q = q_1+q_2.$$
We must therefore show that $p \sin x+q \cos x$ can be put into the equivalent form $r \sin(x+\alpha)$.

As this is a matter of some importance it will be dealt with in a separate section.

5.2 The form $p \sin x+q \cos x$. Since
$$r \sin(x+\alpha) = r \cos\alpha \sin x+r \sin\alpha \cos x,$$
we shall be able to express $p \sin x+q \cos x$ as $r \sin(x+\alpha)$ provided we can find r and α such that $p = r \cos\alpha$ and $q = r \sin\alpha$.

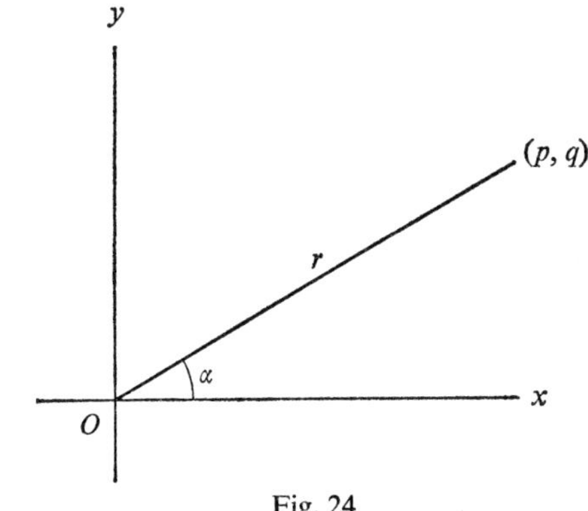

Fig. 24

From the figure it can be seen that this is always possible. In fact, (r, α) are the polar coordinates corresponding to the Cartesian coordinates (p, q). Hence it is always possible to find r, α such that

$$p \sin x + q \cos x = r \sin (x+\alpha).$$

[Alternatively, $p \sin x + q \cos x$ can be expressed in the form $r \cos (x+\beta)$.]

Figure 25 illustrates a geometrical way of obtaining the same results.

The vector **OP** has modulus p and angle x (from Ox); the vector **PQ** has modulus q and angle $x+\frac{1}{2}\pi$; and their sum **OQ** has modulus r and angle $x+\alpha$, where α can be found from triangle OPQ.

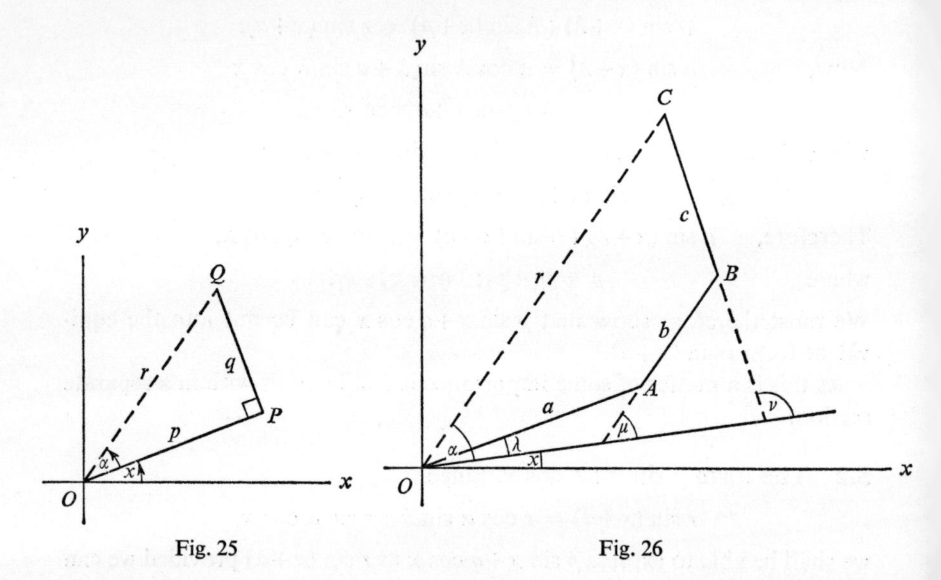

Fig. 25 Fig. 26

By considering the components of these vectors in the y-direction,

$$\begin{aligned} p \sin x + q \cos x &= r \sin (x+\alpha) \\ &= r \cos (x+\alpha-\tfrac{1}{2}\pi). \end{aligned}$$

An extension of the method allows us to express the sum of any number of terms in the form $a \sin (x+\lambda)$ as a single term of the same form.

Figure 26 shows the geometrical method applied to the sum

$$a \sin (x+\lambda) + b \sin (x+\mu) + c \sin (x+\nu).$$

We see how it determines r and α such that the sum equals $r \sin (x+\alpha)$.

The left-hand side is the sum of the y-components of the vectors **OA**, **AB**, **BC**, which is equal to the y-component of their vector sum **OC**, and can therefore be expressed in the form $r \sin (x+\alpha)$.

Example 1. Express $3 \sin \theta - 4 \cos \theta$ in the form $r \sin(\theta+\alpha)$.

Since $$r \sin(\theta+\alpha) = r \cos \alpha \sin \theta + r \sin \alpha \cos \theta,$$

this equals $3 \sin \theta - 4 \cos \theta$ if $r \cos \alpha = 3$ and $r \sin \alpha = -4$. Values for r and α which satisfy these equations are shown in Figure 27, from which it can be seen that $r^2 = 3^2+4^2 = 25$, so that $r = 5$ and

$$\frac{r \sin \alpha}{r \cos \alpha} = \tan \alpha = \frac{-4}{3},$$

so $$\alpha = 2\pi - 0{\cdot}927 = 5{\cdot}356.$$

Hence $$3 \sin \theta - 4 \cos \theta = 5 \sin(\theta + 5{\cdot}356).$$

Since α can equally well be taken as $-0{\cdot}927$, an alternative form is $5 \sin(\theta - 0{\cdot}927)$.

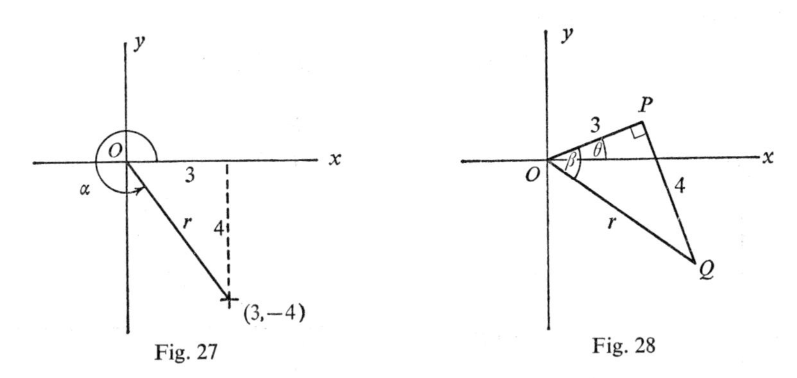

Fig. 27 Fig. 28

Figure 28 shows another way of obtaining these results. For the vector **OP** the y-component is $3 \sin \theta$; and for the vector **PQ** it is $-4 \cos \theta$. By calculation it is found that $r = 5$ and $\beta = 0{\cdot}927$, so the results follow by taking the y-components of the vector equation **OP**+**PQ** = **OQ**.

Example 2. Find the maximum and minimum values of $3 \sin \theta - 4 \cos \theta$, and the corresponding values of θ in the interval 0 to 2π.

From the previous example

$$3 \sin \theta - 4 \cos \theta = 5 \sin(\theta - 0{\cdot}927);$$

so the maximum value is 5 where

$$\theta - 0{\cdot}927 = \tfrac{1}{2}\pi,$$

that is $\theta = 2{\cdot}498$; and the minimum value is -5 where $\theta - 0{\cdot}927 = \tfrac{3}{2}\pi$, that is $\theta = 5{\cdot}639$.

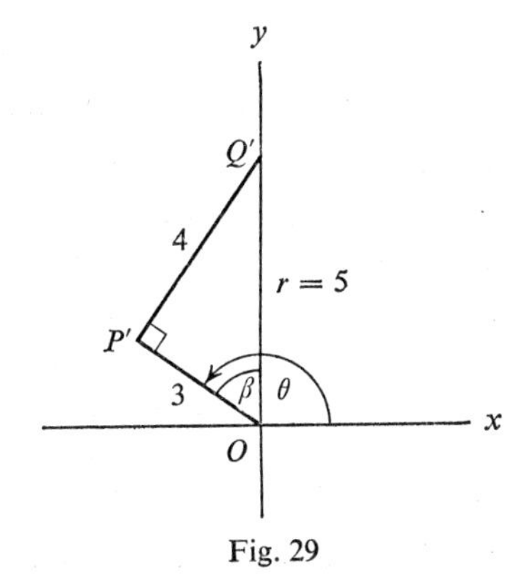

Fig. 29

Alternatively the results can be obtained by considering the triangle OPQ of Figure 28 rotated into triangle $OP'Q'$ of Figure 29. The maximum value is 5 since **OQ** has the greatest y-coordinate when in the position **OQ**$'$, and here $\theta = \frac{1}{2}\pi + \beta = 2{\cdot}498$. Likewise the minimum value -5 is obtained after a further rotation through π, so that $\theta = 5{\cdot}639$.

Example 3. Express $5 \cos \theta + \sin \theta$ in the form $r \cos(\theta - \alpha)$.

Since

$$r \cos(\theta - \alpha) = r \cos \alpha \cos \theta + r \sin \alpha \sin \theta,$$

comparison with $5 \cos \theta + \sin \theta$ gives $r \cos \alpha = 5$, $r \sin \alpha = 1$; and r and α are shown in Figure 30. From this,

$$r = \sqrt{(5^2 + 1^2)} = \sqrt{26},$$

and $\tan \alpha = \frac{1}{5}$, so that $\alpha = 0{\cdot}197$. Hence

$$5 \cos \theta + \sin \theta = \sqrt{26} \cos(\theta - 0{\cdot}197).$$

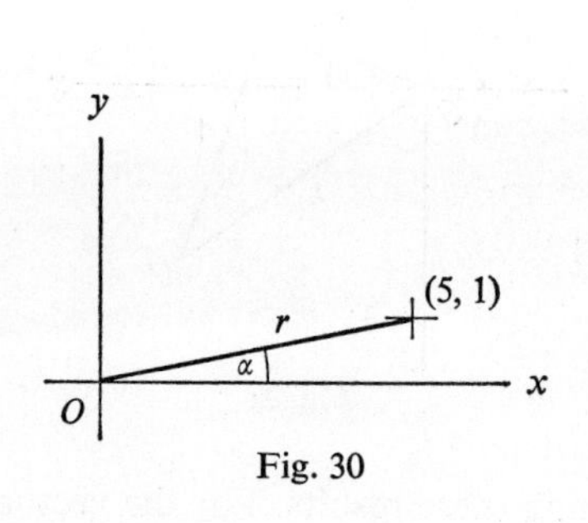

Fig. 30

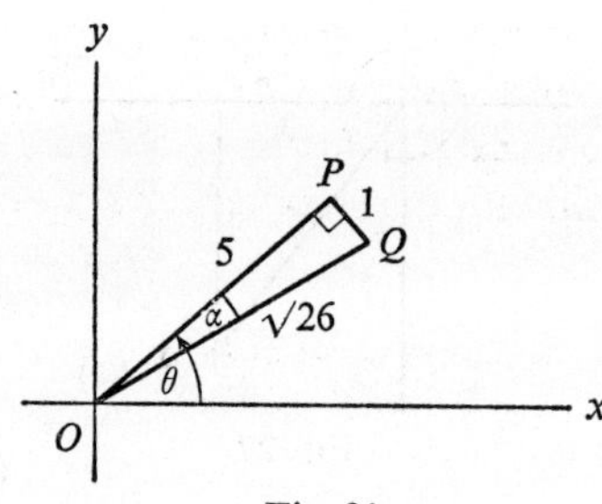

Fig. 31

Alternatively the result follows from Figure 31 by considering the x-components of the vector equation **OP** + **PQ** = **OQ**.

Exercise C

1. Express $\sin x + \cos x$ in the form $r \sin(x + \alpha)$.

2. Express $5 \cos x + 12 \sin x$ in the form $r \cos(x - \alpha)$. Sketch the graph.

3. Find the maximum value, and the values of x for which it occurs, of $7 \cos x - 3 \sin x$, and sketch its graph.

4. Sketch the graph of $5 \sin x + 3 \cos x$, and without calculation estimate the values of r and α if this is the graph of $r \sin(x + \alpha)$.

5. Sketch the graphs of $3 \sin x$ and $2 \sin(x + \frac{1}{3}\pi)$ and of their sum.

6. Using the vector method and measurement, express $3 \sin x + 2 \sin(x + \frac{1}{3}\pi)$ in the form: (*a*) $a \sin(x + \alpha)$; (*b*) $b \cos(x + \beta)$; and check the values of the constants against the graph in Question 5.

7. Using the vector method and measurement, express

$$4 \sin (x+1)+\sin (x+2)-5 \sin (x+3)$$

in the form $r \sin (x+\alpha)$.

5.3 Sum of functions of different periods. The wave formations considered so far are characteristic only of pure tones, and there are very few instruments which actually make these. In practice the fundamental musical note is usually accompanied by *overtones*. For instance, when a violin string is vibrating, the frequency depends on the length of the string; however at the same time that the whole length of the string is vibrating there can be superimposed on these subsidiary vibrations occupying the two halves of the string, and producing at the same time a note, usually of lesser amplitude, an octave higher. This is an overtone. Likewise further overtones can be produced, whose frequencies are those of the natural scale. The actual quality of the note depends on the variety and the strength of these overtones, and it is this which gives each instrument its particular character.

When simple wave forms of different periods are combined, they yield more complicated wave forms. For instance, consider a note and its octave overtone: the composite wave function could have an equation of the form

$$f(x) = a \sin x+b \sin 2x,$$

and for various values of a and b we could plot this wave form.

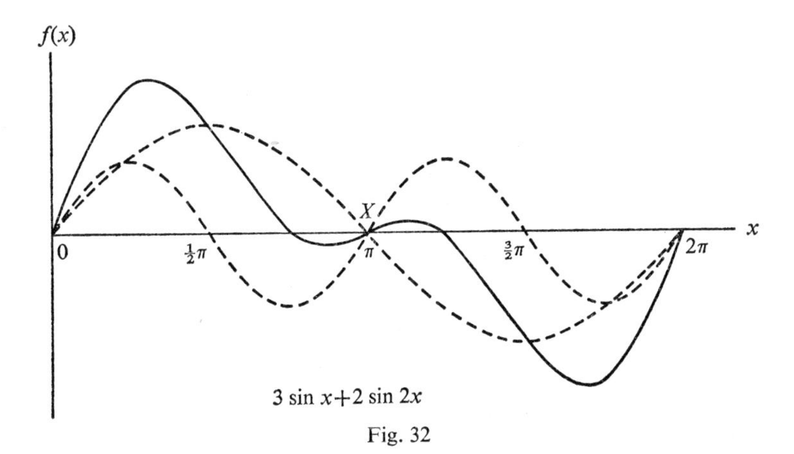

Fig. 32

Figure 32 illustrates the form that it might take. The composite graph will cut that of $a \sin x$ wherever $\sin 2x$ is zero; and from this consideration it is in general easy to sketch. However the form where the two curves cross at $x = \pi$, requires some attention, and depends on the relative sizes of a and b. If the gradient of $b \sin 2x$ at X is steeper than that of $a \sin x$ there will be formed (as shown) subsidiary turning values; and in a particular case we might calculate these. For if the graph is of

$$f(x) = 3 \sin x + 2 \sin 2x,$$
$$f'(x) = 3 \cos x + 4 \cos 2x$$
$$= 8 \cos^2 x + 3 \cos x - 4.$$

Hence
$$f'(x) = 0,$$
where
$$\cos x = (-3 \pm \sqrt{137})/16 \approx \frac{-3 \pm 11{\cdot}7}{16} = 0{\cdot}54 \quad \text{or} \quad -0{\cdot}92.$$

This will give us turning values where $x \approx 1{\cdot}00$, $5{\cdot}29$ or $2{\cdot}73$, $3{\cdot}55$. However if $2b \leqslant a$, this second pair will make no appearance, and there will be a negative gradient inflection at X.

We see the whole graph is that of an oscillating function having the same wave-length as $\sin x$; and this will clearly be true whatever overtones we add, for the wave-length of $\sin x$ is an integral multiple of that of each overtone; thus the modification required in the wave form due to their presence will exactly repeat itself over every cycle of (in this case) 2π. A further point must be added about the phase of the two functions. In the case illustrated we had them both starting at the same point, but this need not have been so. If we change the starting point of one of them we obtain a substantially different graph; for instance, taking

$$a \sin x + b \sin (2x - \tfrac{1}{2}\pi)$$

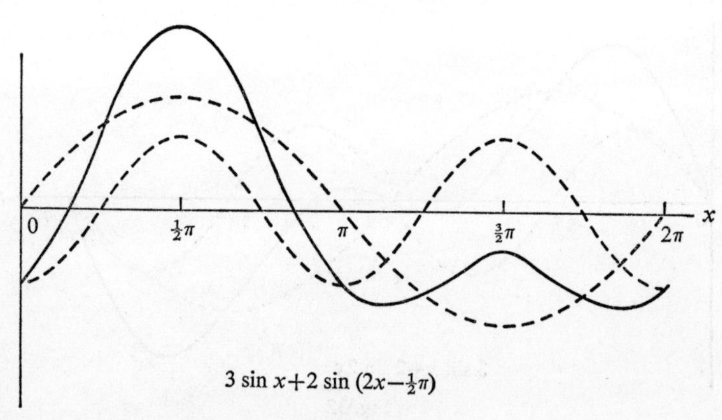

Fig. 33

we obtain the graph in Figure 33. However, although the graph is different, the effect on the ear, which effectively resolves such composite oscillations into their component parts, is exactly the same. Only if we had changed the relative size of a and b would a radically different impression have been created.

Combining a note with its overtones does not produce a new frequency in the composite oscillation. This is not true however when we combine two notes of which one is not an overtone of the other: when for instance two instruments sound different notes at the same time. The common experience in such cases is that although a note combined with another at a fair interval from it in the same scale produces no unpleasant effect—in fact the whole of European music depends on such effects—notes close together, at worst within a tone apart, produce a disagreeable jarring or 'beating' on the ear; and this also can be explained in mathematical terms.

Consider for instance a major third: C and E whose frequencies are in the ratio 4:5. Taking $2 \sin 8x + \sin 10x$ would in effect be adding notes of wave-lengths $\frac{2}{8}\pi$ and $\frac{2}{10}\pi$. Hence we must add the graphs of $2 \sin 8x$ and $\sin 10x$. An alternative method of doing this (which can be useful when we merely want the shape of the curve) is that of taking the 'average' between the components. We see in Figure 34 that the result is a wave formation of wave-length π.

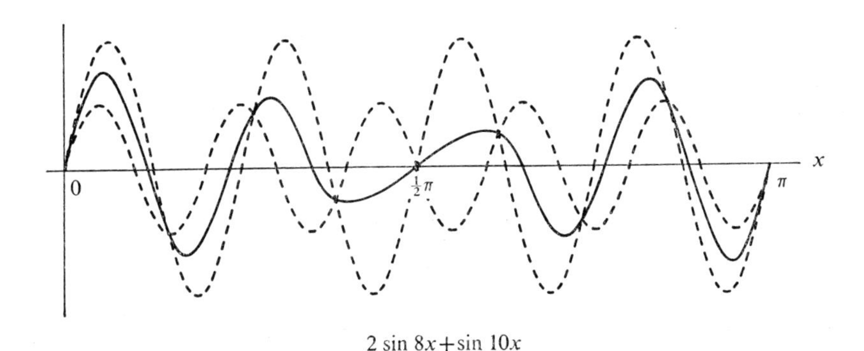

$2 \sin 8x + \sin 10x$

Fig. 34

Compare this now with two notes a tone apart: D and E say, with frequencies in the ratio 9:10. Here we take $\sin 9x + \sin 10x$, now combining notes with wave-length $\frac{2}{9}\pi$ and $\frac{2}{10}\pi$. We get a note of wave-length 2π (see Figure 35): that is to say, if the speed of sound is v, by adding notes with frequency $9v/2\pi$ and $10v/2\pi$ we get one with frequency $v/2\pi$—the difference between the individual frequencies. However this is not the whole of the picture. Within this much larger wave-length, or longer period, there are further oscillations of a period much closer to those of the orginal notes. We get a clear impression of this by analysing the function in a rather different way.

$$\sin 9x + \sin 10x = 2 \sin 9\tfrac{1}{2}x \cos \tfrac{1}{2}x.$$

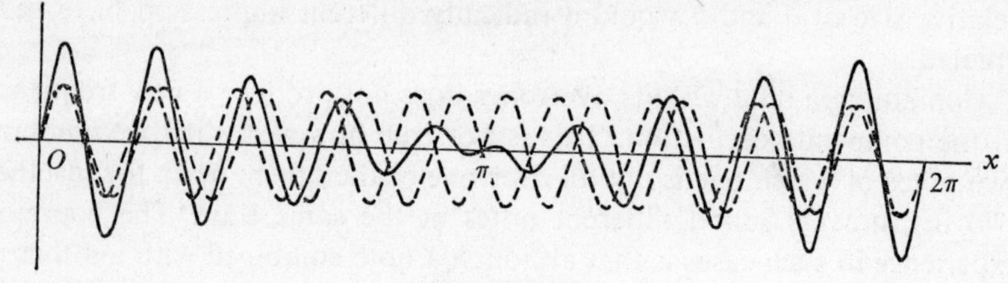

$\sin 9x + \sin 10x$

Fig. 35

We can think then of the wave formation as a rapid oscillation of wave-length $\frac{4}{19}\pi$ (given by $\sin 9\frac{1}{2}x$) bounded by the much slower wave form (given by $2 \cos \frac{1}{2}x$) of period 4π.

Figure 36 shows what, from this point of view, we would expect. The conclusion is that we have a note of somewhere about the same pitch as the original, but that its amplitude oscillates from loud to soft with a much slower frequency. This does not depend on the original notes relative phase,

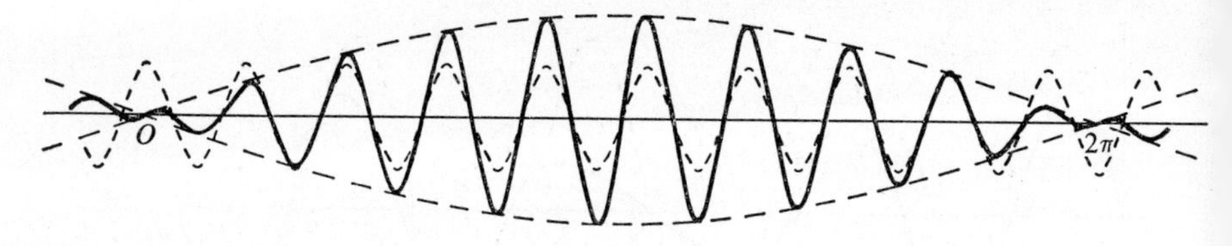

$\sin 9x + \sin 10x$

Fig. 36

for it we turn to Figure 35, we see that it arises because half the time the phase of the two notes reinforce each other, while the other half they cancel each other out. The same effect could have been observed in the previous example, $\sin 8x$ and $\sin 10x$, but here it is slower and more accentuated since the original notes are closer together. This explains the 'beating' caused by two notes at nearly the same pitch. For we know that the loudness of the note depends partly on its amplitude. Hence the oscillation between loud and soft, slow enough to be noticed, but not so slow as to be inoffensive, will irritate the ear rather as a flicker of light will irritate the eye. If the notes indeed are very close together the alternation will be too slow to be objectionable; and if they are farther apart it will be too fast to be noticed, just as the eye does not note the flicker of a railway carriage wheel when it is going faster than a certain speed. There is however a critical range where the impression will be disagreeable.

Exercise D

1. Sketch the graph of $\cos 2x - \sin x$, finding turning values.

2. Sketch the graphs of $\sin 2x$ and $-2 \cos x$, noting their gradients where they cross. Hence, without further use of calculus, sketch the graph of $\sin 2x - 2 \cos x$. Check by calculating turning values. Sketch also, without calculating turning values, the graphs of $\sin 2x - \frac{3}{2} \cos x$ and $\sin 2x - 3 \cos x$. What symmetries have these graphs?

3. Sketch, without calculating turning values, the graph of $\sin 2x + \frac{3}{4} \sin 3x$.

4. Sketch, calculating turning values, the graphs of:

(i) $2 \sin 2x + 5 \sin x$,

(ii) $2 \sin 2x + 3 \cos x$,

(iii) $2 \sin 3x - 3 \cos (2x + \frac{1}{2}\pi)$.

5. At a certain port the tide is due partly to the moon and partly to the sun, and it may be assumed that each part is of a simple sine wave form. If the moon's tide has a maximum height of 2 m, above mean tide level, at 6 a.m. and period 25 hours, while the sun's tide has a maximum height of 1 m at 5 a.m. and a period of 12 hours, find, by sketching the two wave forms and adding them together, the form of the composite tide during the day. (Use the fact that $\sin 30° = \frac{1}{2}$ to plot the heights of the two tides at times equal to a twelfth of the period from the times when they are at mean sea level.)

6. Sketch the wave form of the note C combined with E, that is with frequencies in the ratio 4:5, in two cases to show the difference of wave form that can occur with different relative phase.

7. Show the wave formation of a note combined with its fifth overtone, that is, with that having frequency 6 times as great, where the overtone has very much smaller amplitude than the original note.

8. Show the wave formation of C and D sounded together, that is, notes with frequency in the ratio 8:9, where they have the same amplitude.

9. Compare the result of Question 8 with that of combining a note with one a semitone higher (for example, combining B and C, having frequencies in the ratio 15:16).

10. Sketch the graph of $b \sin x - a \sin^3 x$, $a > b > 0$, calculating maxima and minima. Find the values of a and b for which the maxima and minima are all equal to 1 and -1. Hence find a formula for $\sin 3x$. Does your work amount to a proof of this formula?

11. Sketch the graph of $\sin 2x - 10 \cos^3 x$.

6. PHYSICAL PROBLEMS

6.1 Geometrical applications. In the previous section we emphasized the use of trigonometrical functions in some applications that took no account of their geometrical properties. There are however, as we would expect, a great many geometrical problems where we can use them in their more familiar role. In this section we shall deal with some of these. The reader will observe that in this kind of work trigonometrical functions have certain advantages; they facilitate the use of a single variable, and avoid troublesome square roots incident upon a use of Pythagoras' Theorem. Indeed many of the problems in Chapters 4 and 6 could have as easily been done, and some more easily done, with the aid of trigonometry; hence the repetitive nature of some of the work which follows, although we shall give also examples which could not have been solved before.

Example 1. What is the smallest clôche which will just cover a cylindrical vegetable marrow?

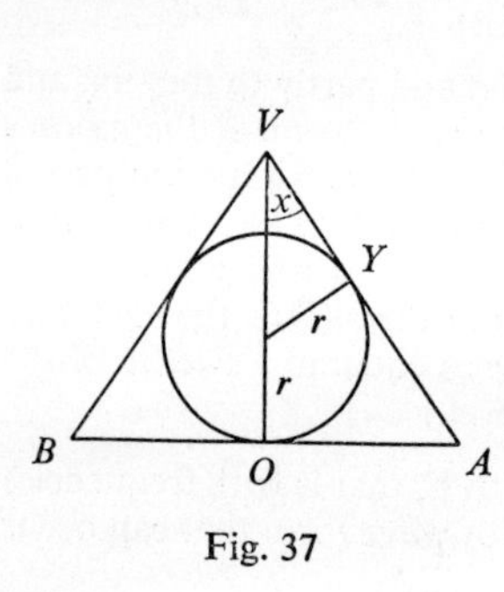

Fig. 37

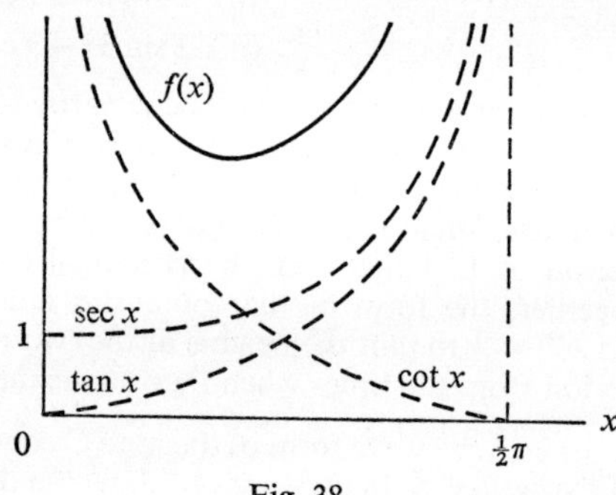

Fig. 38

Here we must, in effect, find the isosceles triangle with the smallest side VA which will enclose a given circle. This might easily have been done by letting the half-base OA be x, and finding the length VA in terms of x and the radius, by an examination of similar triangles. Equally however we can take as variable the *angle* AVO and call it x. Then the height of the triangle is $r(\operatorname{cosec} x+1)$, and hence OA (and so AY) is equal to

$$r(\operatorname{cosec} x+1)\tan x = r(\sec x+\tan x);$$

while VY is $r\cot x$. Hence the function whose minimum value we are to find is

$$f(x) = \sec x+\tan x+\cot x,$$

whose graph is shown.

$$f'(x) = \sec x\tan x+\sec^2 x-\operatorname{cosec}^2 x = s/c^2+1/c^2-1/s^2$$

$$\text{where } s = \sin x,\ c = \cos x$$

$$= (s^3+s^2-c^2)/s^2c^2 = (s^3+2s^2-1)/s^2c^2 = (s+1)(s^2+s-1)/s^2c^2.$$

Hence $f'(x) = 0$, where $s = -1$ or $s = (-1\pm\sqrt{5})/2$. Clearly it is the root $\sin x = (-1+\sqrt{5})/2$ which is relevant, which gives us the shape of cloche which ought to be used.

Example 2. A liner is steaming along a straight track with speed V km/hour. What is the least speed needed by a launch 4 km from the track and 5 km from the liner, in order to intercept her?

Let us suppose here that the launch sets out on a course making an angle x with the track. Then we can calculate the speed needed for it to intercept the liner, and hence find for what value of x this speed is least.

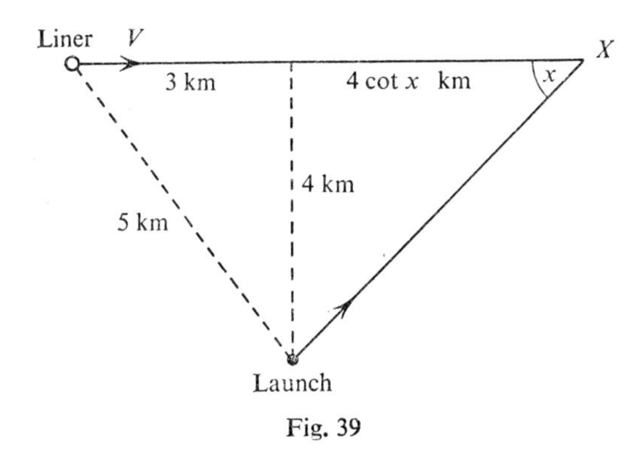

Fig. 39

Now the launch intercepts the liner at X; and the time taken for the liner to get there is $(3+4 \cot x)/V$ hours. In that time the launch must go $4 \operatorname{cosec} x$ km. Hence the speed it requires is

$$4V \operatorname{cosec} x/(3+4 \cot x) = \frac{4V}{3 \sin x + 4 \cos x}.$$

We do not need calculus to find the minimum of this function. We can express it in the form

$$4V/[5 \sin (x+\alpha)], \quad \text{where} \quad \tan \alpha = \tfrac{4}{3}.$$

This has minimum value when $\sin (x+\alpha)$ is maximum, that is, when $x+\alpha = \frac{1}{2}\pi$. Hence the least velocity needed is $\frac{4}{5}V$ km/hour, and the angle at which the launch should go is $\frac{1}{2}\pi - \tan^{-1}(\frac{4}{3})$ to the liner's course. This is at right-angles to the line joining its initial position to that of the liner.

**Example 3.* Figure 40 shows a reel mechanism for playing a fish. Fixed to the rod handle is a geared wheel centre O of radius a cm. Another geared wheel with centre A, of the same radius, engages the fixed wheel and can be moved so that A described a circle with centre O. The line is wound onto a spindle of radius b cm, concentric and combined with the moveable wheel. Describe the motion of the fish as it is hauled in.

We shall assume that the rod is long enough for it to be a reasonable approximation that the line runs from the reel in a direction parallel to the rod. It will be convenient also at first to neglect the increase in the effective radius of the reel as the line is wound on.

Let AO make an angle θ with the rod as shown; and let the fish be at a distance d cm from O (measured along the line) when $\theta = 0$. When OA turns through angle θ the point X where the line leaves the reel has moved a distance $2a(1-\cos\theta)$ cm 'inwards'—the same as the distance moved by A.

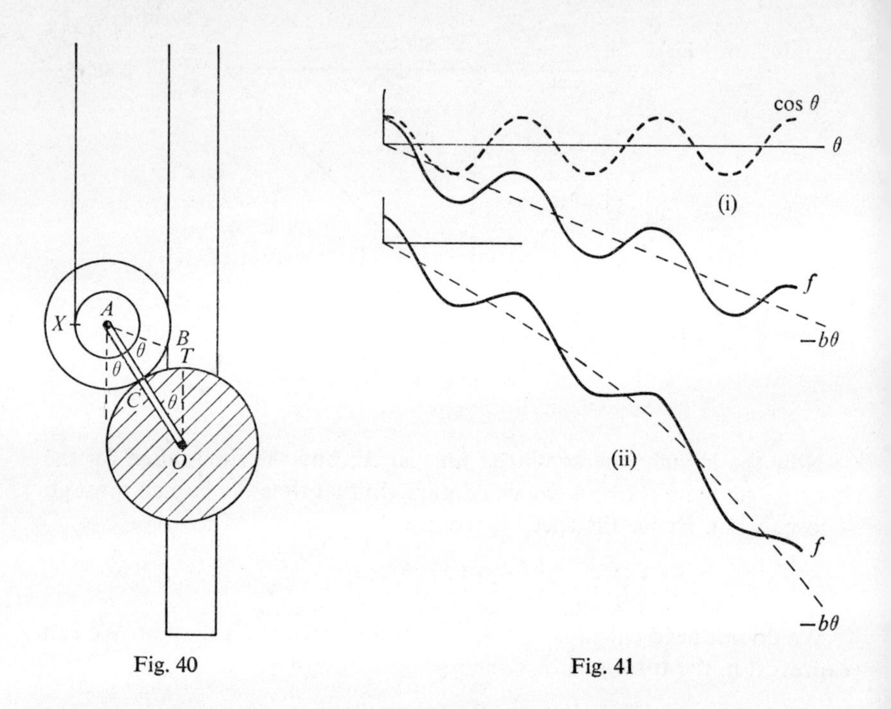

Fig. 40 Fig. 41

In this motion the moveable wheel has turned through 2θ, because if B originally coincided with T, $\angle BAC = \angle TOC$, so that the radius AB has moved from the 'vertical' through 2θ. A length of $2b\theta$ cm has therefore been wound onto the reel. If the fish is now at a distance x cm from O,

$$x = d-2a(1-\cos\theta)-2b\theta$$
$$= (d-2a)+2(a\cos\theta-b\theta).$$

The graph of $f(\theta) = a\cos\theta-b\theta$ is shown in Figure 41 (i) for $b < a$. Now $f'(\theta) = -a\sin\theta-b = 0$ if $\theta = \sin^{-1}(-b/a)$. These values of θ give the two positions each cycle where the fish is stationary: once when about to come closer and once when about to be paid out. From the graphical addition it is clear that these values of θ occur before the maximum and after the minimum positions for $\cos\theta$.

To allow for the winding on the reel growing fatter as the line is wound in, we need to consider the effect of b increasing with θ. Instead of the straight-line graph for $b\theta$ we now have a graph which curves downwards as in Figure 41 (ii). The graph of $f(\theta)$ will continue to oscillate about the graph of $b\theta$ but as the latter becomes steeper $f(\theta)$ will change to a function which is always decreasing (although unsteadily). This happens as soon as b exceeds a. The fish at this stage is always being hauled in although at a variable speed.

Exercise E

1. A cylindrical cheese is to be cut out of a spherical one. Express its volume as a function of half the angle subtended by a diameter of its circular end at the centre of the sphere. Hence show that the biggest cheese that can be cut out in this way is $(\frac{1}{3})^{\frac{1}{2}}$ of the sphere.

2. P is a point outside a circle of centre O, where $OP = 13$ cm. Tangents from P to the circle touch it at A and B. For what value of the angle OPA is the area of $\triangle PAB$ greatest?

3. A psychologically deranged Eskimo wants to conceal a hemispherical igloo (of radius 3 m) beneath a conical wigwam. By expressing the surface area of the wigwam as a function of its semi-vertical angle, and considering the reciprocal of this function, or otherwise, find the least amount of sealskin he needs to do this.

4. A hemispherical soup tureen has diameter 24 cm. A ladle whose lower end is in line with its straight shaft, leans with that end against the inner surface of the tureen, and its shaft against the edge. P is a point on the shaft, 4 cm from its lower end. Express the depth of P below the top of the tureen as a function of the angle the ladle makes with the horizontal. Is it possible, if there are 3 cm of soup in the tureen, for this point to get wet?

5. A man is to swim across a river 50 units wide with uniform current 5 units/s. If he can swim at 3 units/s in still water, find, by considering the components of his velocity parallel to and at right angles to the banks, what is the highest point he can reach on the other side.

6. A tank can be driven along the bank of a river at 5 m/s, and over the river bed at 3 m/s. The driver wants to cross the river (which is 25 m wide) to a point 20 m downstream. At what angle to the bank should he go, if he is to cross as quickly as possible? Would the answer be different if the point were 15 m downstream?

7. Find the biggest right circular cone that can be fitted inside a sphere of radius r.

8. A man has a 13·5 m ladder in a narrow corridor 4 m high. He has wagered with his wife that he can reverse the ends of the ladder without taking any part of it beyond the ends of the corridor. Unknown to his wife there is a small trapdoor in the ceiling leading to an attic 10 m high. If the door is in the middle of the attic floor which is 11 m wide, who wins?

9. A man wishes to get from A to B where AB is the diameter of a circular goldfish pond. He can run round the edge k times faster than he can wade. Discuss, for various ranges of values of k, his quickest way of doing it.

10. A missile is homing on its target with a speed of 1000 m/s. An anti-missile missile is launched from a position 400 m from its track and is capable of a speed of 800 m/s. Assuming that it intercepts at an angle θ to the track, express as a function of θ the distance between the missiles parallel to the track at the instant of launching the second. Show that if this distance is 320 m there are two possible lines of interception. What is the least possible distance for interception?

11. A wing three-quarter is 26 m from the opposing goal line running down the touch line at 8 m/s. The opposing fullback is 12 m in from the touch line and 6 m from his own goal line. Express the speed at which he must run in order to intercept, as a function of the angle which his course makes with the touch line. Hence find the minimum speed he needs in order to prevent a try.

12*. Discuss the solubility of the equation $\tan x = kx$, in the range $-\frac{1}{2}\pi$ to $+\frac{1}{2}\pi$ for different ranges of values of k. A see-saw consists of a plank of thickness d, which rolls without slipping on a fixed pipe of radius a, whose axis is horizontal. When the plank is horizontal, P is a point on top of it vertically above the axis. Express the height of P above the axis of the pipe as a function of the angle the plank makes with the horizontal. Hence discuss the highest points attained by P (i) when $a > d$, (ii) when $a < d$, and (iii) when $a = d$.

13*. A metre rule has its ends moving in perpendicular grooves OA and OB. A fixed line parallel to OB at a distance d (< 1 m) from it, cuts OA at K. If it cuts the rule at X, express KX as a function of the angle the rule makes with AO. Hence show that the plane OAB is divided into points through which the rule passes twice, once, or not at all. Find the locus of the points through which it passes once and sketch its form.

14*. 20 m^3 of water are in a rectangular container measuring 5 m by 5 m by 4 m high. If it is slowly tipped about one edge of the base, express the area of the surface of the water as a function of the angle through which it is tipped, and sketch its graph.

15. Cotton is wound off a reel at 1 revolution per second in a plane at right-angles to its axis. Taking the origin of coordinates on the axis and the free end of the cotton to lie originally at the point $(a, 0)$ where a is the radius of the reel, express the coordinates of the end as functions of time. Hence find the velocity of the end and comment on your result.

16. A small flint is wedged in the tread of a car tyre of radius r. If initially it is in contact with the road, express (i) the vertical distance, and (ii) the horizontal distance it has moved as a function of the angle through which the tyre has turned. Sketch the graph of these functions. Hence find in what direction the flint is moving initially and then sketch the locus of the flint.

17*. A lamp, the horizontal cross-section of whose arc of light makes an angle of $\frac{1}{2}\pi$, is 6 m from the centre of one wall of a square room of side 15 m. Sketch the graph of the horizontal length of wall illuminated as a function of the angle through which it has turned from its original position facing perpendicularly on to the nearest wall.

18*. The following is a 'mathematical model' of a sailing ship and its sail. The sail is a flat smooth surface, which can be set at any angle to the axis of the ship. The force from the wind on the sail, for constant wind, is thus at right-angles to the sail; it is further proportional to the component of the wind's velocity at right-angles to the surface of the sail. The action of the ship's rudder and keel are such that the ship can only move in the line of its own axis; the resistance to the ship in the direction of its axis from the water (which is dead calm with no tide flowing) is proportional to the ship's speed relative to the water.

(*a*) Show that if the wind makes an angle x with the course of the ship, then the sail is best set so that it bisects the angle between the ship's course and the wind; and that the velocity of the ship is $kw(1+\cos x)$, where w is the wind's speed and k some constant.

(*b*) Use this result to show that if the ship is trying to move dead against the wind, its best course is to tack on paths inclined at 60° to the direction in which it is to move.

What are the *main* objections to this mathematical model?

10

EXPANSION IN SERIES

We are familiar with the idea that a linear approximation to a function at a point is an approximation to the function near that point. It is reasonable to suppose that there could be non-linear approximations which by incorporating some of the non-linearity of the function would provide better approximations. This is the problem we are about to study.

1. NON-LINEAR APPROXIMATIONS

1.1 Approximations to polynomial functions. We start by considering approximations to a simple polynomial function f, where

$$f(x) = 12+3x+2x^2-x^3.$$

Near $x = 0$, each term of this cubic is small compared with its predecessor, so we would expect 12, $12+3x$, $12+3x+2x^2$ all to give approximate values to $f(x)$; but $12+3x$ better than 12, and $12+3x+2x^2$ better than $12+3x$.

Let us consider the graphical relation between the two approximations $12+3x$ and $12+3x+2x^2$, and $f(x)$ itself near $x = 0$. Since f_1 is the linear approximation at $x = 0$ to f_2, its graph touches that of f_2 at P (see Figure 1). The graph of f_2 can be drawn by adding that of $2x^2$ to the graph of f_1, and therefore lies wholly above the graph of f_1 (except at P). How then does the graph of f lie in relation to those of f_1 and f_2? It touches f_1 at P because f_1 is its linear approximation there. It can be drawn by adding $-x^3$ to the graph of f_2, and therefore when x is negative it is above that of f_2. Now when x is positive the graph of f is below that of f_2 by the amount x^3 (HG), whereas the graph of f_1 is below that of f_2 by the amount $2x^2$ (HK); so, near $x = 0$, for positive values of x, since x^3 is small compared with $2x^2$, the graph of f will lie between those of f_1 and f_2. But we can say more than this: x^3 (HG) is small compared with $2x^2$ (HK) implies that x^3 (HG) is small compared with $2x^2-x^3$ (GK); hence, near $x = 0$, we are justified in saying that f_2 is a closer approximation to f than is f_1. This holds also for negative values of x, but in this case the result is obvious because the graph of f_1 is below that of f_2.

We must be careful to notice that although within some interval the quadratic approximation clings more closely than the linear one to the parent curve, this is not necessarily true over all intervals. In this example, if x is negative the quadratic is always the better approximation; and is

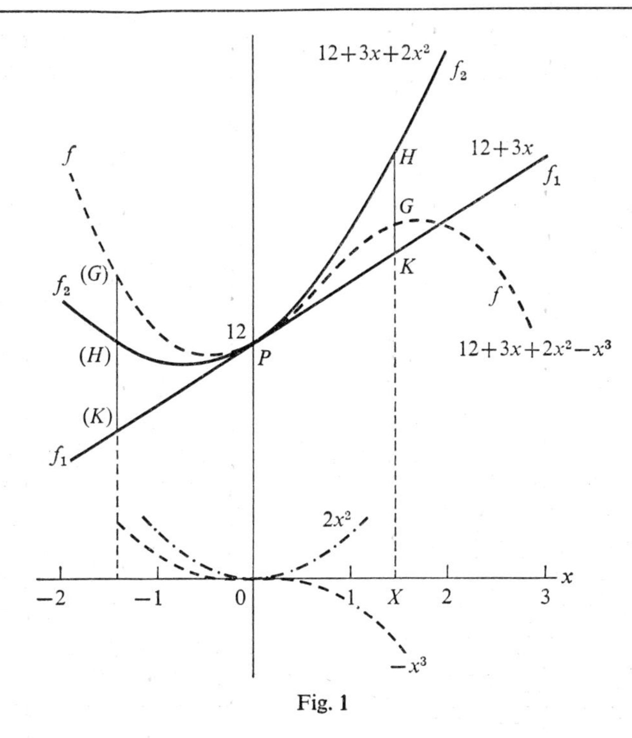

Fig. 1

better if x is positive so long as $x^3 < 2x^2 - x^3$, that is, for $x < 1$; but for $x > 1$, the linear approximation is the better (for what it is worth).

It should also be realized that we set out to consider approximations near a point. If instead we had been concerned to find, for instance, a suitable linear approximation to f over a specific interval, say $-1 < x < 1$, we might well have preferred one whose graph is l (see Figure 2) to that of f_1, since the errors are 'on the whole' smaller. Quite what should govern our preference is a matter on which the reader might like to ponder (see Exercise A, Questions 7, 8).

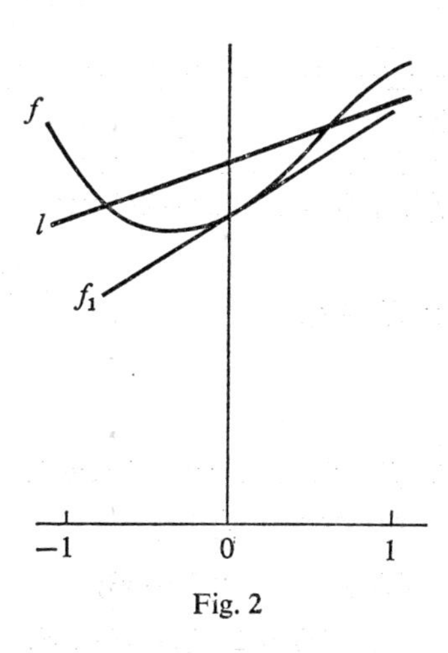

Fig. 2

The distinction between these two types of approximation is emphasized here because in the rest of this chapter we shall restrict attention to approximations near a point, and it would be a pity if the reader were to form the impression that these are the only kind or indeed the only important kind of approximation. There are many occasions in applied mathematics when an approximation over a specific interval is more desirable.

When we obtain a polynomial approximation $P(x)$ to a function of x, $f(x)$, near a point, we shall require that the numerical difference between $P(x)$ and $f(x)$ can be made as small as we like throughout *some* interval of the domain which includes the point; but the actual extent of this interval will depend in each case on how small is the difference sought.

There are some further ideas we can extract from the example with which we started before we leave it to look at one where an approximation is found to a non-polynomial function. First we ask: what non-linear feature of f is incorporated in the approximation f_2?

For any value of $x, f'(x)$, which is $3+4x-3x^2$, measures the gradient of the graph at the corresponding point; the derivative of $f'(x)$, which is $4-6x$, will then measure the rate of change of this gradient, and we shall write this as $f''(x)$. Hence $f''(x)$ is a measure of the degree of bending of the curve at that point. (It is not the ideal measure of this property because a given change of gradient over a certain interval for x corresponds to a sharper bend if the gradient is small than if it is large.) Now the linear approximation f_1 is such that $f_1(0) = f(0)$ and $f_1'(0) = f'(0)$; that is, the graphs of f and f_1 pass through the same point P with the same gradient; but the quadratic approximation has the additional property that

$$f_2''(0) = f''(0),$$

since $f_2''(x) = 4$ and $f''(x) = 4-6x$. Hence the graph of f_2 not only passes through the same point P with the same gradient as f but they have the same rate of change of gradient there.

A further point to notice is that there is an obvious extension to approximations in the neighbourhoods of values of x other than zero. If instead we had known that $f(p+\epsilon) = 12+3\epsilon+2\epsilon^2-\epsilon^3$, then the same argument as we had earlier would have shown that near $x = p$, the quadratic approximation $f_2(p+\epsilon) = 12+3\epsilon+2\epsilon^2$ was better than the linear approximation

$$f_1(p+\epsilon) = 12+3\epsilon,$$

to $f(p+\epsilon)$.

Exercise A

1. Sketch in the same figure the graphs of:

(i) $f_1(x) = 2+\frac{1}{2}x$;

(ii) $f_2(x) = 2+\frac{1}{2}x-x^2$;

(iii) $f_3(x) = 2+\frac{1}{2}x-x^2+2x^3$.

2. For each of the following functions of x, state its successive approximations near $x = 0$, and show how the graphs of these lie in relation to that of the function itself near $x = 0$:

(i) $f(x) = -3-x-x^2+2x^4$;

(ii) $f(x) = 2x+3x^3-x^4$;

(iii) $f(x) = 5+2x^2-x^3$.

3. $f(x) = 1-2x-x^3-x^4$. Within what interval (roughly) does (i) the first approximation, and (ii) the second approximation give a value for $f(x)$ correct to five decimal places? In what interval is the second approximation better than the first? Sketch their graphs together, in the interval $(-2, 2)$

4. Sketch the graphs of the following functions f near $x = -3$, if it is known that $f(-3+\epsilon)$ is:

(i) $2+\epsilon-3\epsilon^2$; (ii) $2-\epsilon+\epsilon^3$;

(iii) $\epsilon+\epsilon^2-2\epsilon^5$; (iv) $3-\epsilon^3$.

5. Form the successive approximations to $f(x)$ near $x = 10$, if it is known that

$$f(10+\epsilon) = -6+\tfrac{1}{2}\epsilon-2\epsilon^2+\epsilon^3,$$

and sketch their graphs.

6. It is known that $f(2+\epsilon) = 0{\cdot}1-0{\cdot}2\epsilon-\epsilon^3$.
Sketch the graph in the interval $(1, 3)$ and hence estimate, to 1 decimal place, a root of the equation $f(x) = 0$.

7. Find a linear approximation to $f(x) = x^2$ over the interval $0 \leqslant x \leqslant 2$, in the following way. Suppose the approximation is given by $g(x) = a+bx$; then choose a, b so that the largest value of $|f(x)-g(x)|$ at the three points where $x = 0, 1, 2$ is as small as possible.

8. An alternative method to that of the previous question is as follows. Show that the sum of $[f(x)-g(x)]^2$ at the three points is $3a^2+5b^2+6ab-10a-18b+17$. If b is fixed but a varies, show that the least value of the sum occurs when $3a+3b = 5$; and if a is fixed but b varies, it occurs when $3a+5b = 9$. Find the linear approximation which satisfies both these conditions. (It can be shown that this does in fact give the least value for the sum.)

1.2 Approximations to non-polynomial functions. A particularly good illustration of the way in which a non-polynomial function can be approximated to by a polynomial function is afforded by the sine function. We shall obtain a sequence of polynomial approximations near $x = 0$, of increasing degree, which have the property that if we choose arbitrarily any interval of the domain, and specify what non-zero error is acceptable, then there are polynomials of this sequence which approximate to $\sin x$ within this degree of accuracy over the interval. This is not a general property of polynomial approximations, but it makes an interesting first example.

It is convenient to start by restricting x to be non-negative; then we can base our analysis on the simple fact that if f is a continuous function which is zero when x is zero and which has a positive derivative (or occasionally a zero derivative) for all $x > 0$, then $f(x) > 0$ for all $x > 0$. This will no doubt strike the reader as obvious and we shall not attempt any justification.

Now consider $s_1(x) = x - \sin x$.

$s_1'(x) = 1 - \cos x$, so that $s_1'(x) > 0$ if $x > 0$ (except where $\cos x = 1$); also $s_1(0) = 0$. Hence

$$s_1(x) = x - \sin x > 0 \quad \text{if} \quad x > 0.$$

It follows that if we now construct a function s_2 whose derivative is s_1 and for which $s_2(0) = 0$, this also must always be positive. By integrating $x - \sin x$ we have

$$s_2(x) = \tfrac{1}{2}x^2 + \cos x + \text{constant},$$

and as $s_2(0) = 0$, the constant is -1. Hence

$$s_2(x) = \tfrac{1}{2}x^2 - 1 + \cos x > 0 \quad \text{if} \quad x > 0.$$

Again we construct another function s_3 whose derivative is s_2 and for which $s_3(0) = 0$. This also must always be positive. By integrating $s_2(x)$ and finding the constant so that $s_2(0) = 0$, we find

$$s_3(x) = \frac{x^3}{2.3} - x + \sin x > 0 \quad \text{if} \quad x > 0.$$

Similarly $$s_4(x) = \frac{x^4}{2.3.4} - \frac{x^2}{2} + 1 - \cos x > 0 \quad \text{if} \quad x > 0,$$

and so on.

Selecting the results for s_1, s_3, s_5, s_7 and so on, we have†

$$x > \sin x; \tag{i}$$

$$\sin x > x - \frac{x^3}{3!}; \tag{ii}$$

$$x - \frac{x^3}{3!} + \frac{x^5}{5!} > \sin x; \tag{iii}$$

$$\sin x > x - \frac{x^3}{3!} + \frac{x^5}{5!} - \frac{x^7}{7!} \tag{iv}$$

and so on.

From (i) and (ii) we have $-\frac{x^3}{3!} < \sin x - x < 0$, which shows that x is an approximation to $\sin x$ where the error is less than $\left|\frac{x^3}{3!}\right|$, which is small compared with x. This is the linear approximation at $x = 0$. It holds also for negative values of x, because replacing x by $-x$ merely changes the sign of $\sin x$, x and x^3; so there is a corresponding inequality statement with $<$ replaced by $>$.

† It is an accepted notation to write the product $1.2.3 \ldots n$ as $n!$; it is called 'n factorial'.

A quadratic approximation does not arise and would have been an embarrassment when extending to negative values of x because of the point symmetry of the graph of $\sin x$ about the origin. But if we combine (ii) and (iii) we get $0 < \sin x - \left(x - \frac{x^3}{3!}\right) < \frac{x^5}{5!}$ and obtain the next approximation to $\sin x$ as $x - \frac{x^3}{3!}$ with an error less than $\left|\frac{x^5}{5!}\right|$ which is small compared with x^3.

Similarly we can derive further approximations. We shall call the successive approximations to $\sin x$: $f_1(x), f_3(x), f_5(x)$, and so on. They are:

$$f_1(x) = x,$$

$$f_3(x) = x - \frac{x^3}{3!},$$

$$f_5(x) = x - \frac{x^3}{3!} + \frac{x^5}{5!}, \quad \text{and so on.}$$

All of these are equally valid for negative values of x for reasons similar to those given earlier for the linear approximation.

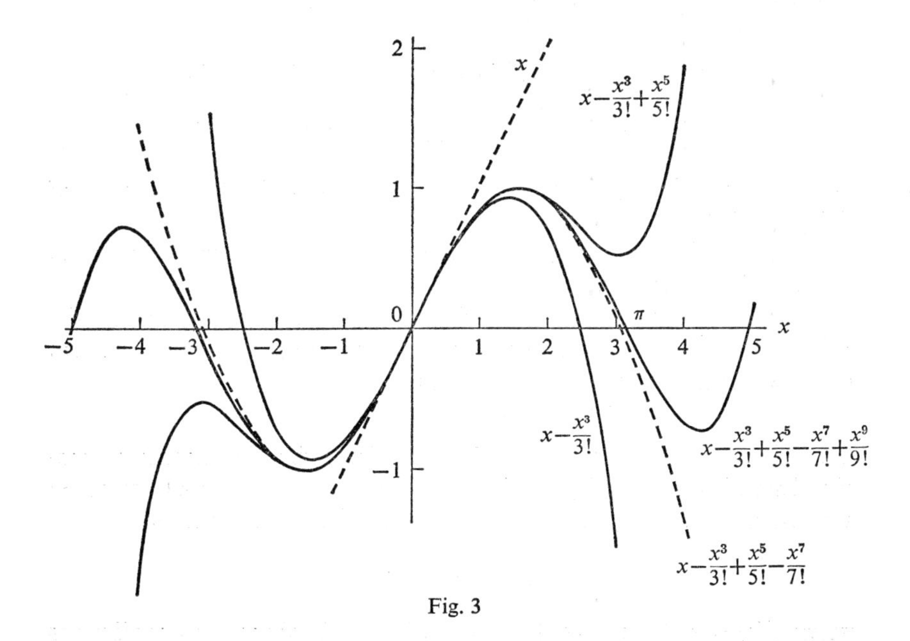

Fig. 3

It is interesting to see the characteristic sine curve shape emerging from the successive approximations. This is indicated in Figure 3 which shows the graphs of f_1, f_3, f_5, f_7, f_9.

Using the approximation $\sin x \approx x - \frac{x^3}{3!} + \ldots - \frac{x^{19}}{19!}$ for which the error is less than $\left|\frac{x^{21}}{21!}\right|$, it can be shown that for $x = 2\pi$, $(2\pi)^{21} = 5{\cdot}78 \times 10^{16}$ and $21! = 5{\cdot}11 \times 10^{19}$; so there is close agreement between $\sin x$ and this approximation over the range -2π to $+2\pi$ (two complete waves) the greatest error being about $1{\cdot}1 \times 10^{-3}$. If the approximation is taken as far as $\frac{x^{99}}{99!}$, the error over the range -10π to $+10\pi$ is less than $1{\cdot}7 \times 10^{-9}$ and over the range -12π to $+12\pi$ less than $1{\cdot}7 \times 10^{-1}$. If we take n terms in the approximation we can, by making n large enough, get as close as we like to $\sin x$, over any chosen range, because the error term $\left|\frac{x^{2n+1}}{(2n+1)!}\right|$ can then be made as small as we please (see Exercise B, Question 10). This, it must be repeated, is not a general feature of approximations; it frequently happens that, however many terms are taken, the approximation is only valid over a certain range.

Let us now look at an important attribute of these approximations. For $s: x \to \sin x$ and $f_5: x \to x - \frac{x^3}{3!} + \frac{x^5}{5!}$, it is easily checked that

$$s(0) = f_5(0), \qquad s'(0) = f_5'(0), \qquad s''(0) = f_5''(0),$$

$$s'''(0) = f_5'''(0), \quad s^{(iv)}(0) = f_5^{(iv)}(0), \quad s^{(v)}(0) = f_5^{(v)}(0)$$

(where for any function ϕ we use the notation $\phi'''(x)$ for the derivative of $\phi''(x)$, $\phi^{(iv)}(x)$ for the derivative of $\phi'''(x)$, and so on). In general, for approximations of degree n, their first n successive derivatives are equal to those of $\sin x$ at $x = 0$.

This suggests a general method for finding approximations in the neighbourhood of a point to any function whose successive derivatives exist there; this is developed in the next section.

1.3 Taylor's approximation. Suppose f is a function having successive derivatives $f'(a), f''(a), \ldots, f^{(n)}(a)$, at $x = a$. We look for a polynomial approximation to $f(x)$ in the neighbourhood of $x = a$. Let it be

$$f(a+\epsilon) \approx \phi(\epsilon),$$

where $\phi(\epsilon) = p_0 + p_1\epsilon + p_2\epsilon^2 + \ldots + p_n\epsilon^n$. We shall choose the coefficients $p_0, p_1, \ldots, p_n$, so that $f(a+\epsilon)$ and $\phi(\epsilon)$ are equal and have the same successive derivatives when $\epsilon = 0$, as far as the nth derivative.

Putting

$$\phi(0) = f(a) \quad \text{we see that} \quad p_0 = f(a).$$

Now $\phi'(\epsilon) = p_1+2p_2\epsilon+3p_3\epsilon^2+ 4p_4\epsilon^3+\ldots+np_n\epsilon^{n-1};$

$$\phi''(\epsilon) = 2p_2+3.2p_3\epsilon+4.3p_4\epsilon^2+\ldots+n(n-1)\,p_n\epsilon^{n-2};$$

$$\phi'''(\epsilon) = 3.2p_3+4.3.2p_4\epsilon+5.4.3p_5\epsilon^2+\ldots+n(n-1)\,(n-2)\,p_n\epsilon^{n-3};$$

and so on.

Putting $\phi'(0) = f'(a)$ we see that $p_1 = f'(a)$;

and $\phi''(0) = f''(a)$ we see that $p_2 = \frac{1}{2}f''(a)$;

and $\phi'''(0) = f'''(a)$ we see that $p_3 = \dfrac{1}{3.2}f'''(a)$;

and, in general,

$\phi^{(r)}(0) = f^{(r)}(a)$ we see that $p_r = \dfrac{1}{r!}f^{(r)}(a)$.

Hence, for our approximation, we have

$$f(a+\epsilon) \approx f(a)+f'(a)\,\epsilon+\frac{1}{2!}f''(a)\,\epsilon^2+\ldots+\frac{1}{n!}f^{(n)}(a)\,\epsilon^n.$$

This is known as Taylor's approximation.

As we would expect, if we apply this to the sine function near $x = 0$ with a, ϵ replaced by 0, x, so that

$$\sin x \approx \sin(0)+\sin'(0)\,x+\frac{1}{2!}\sin''(0)\,x^2+\ldots+\frac{1}{n!}\sin^{(n)}(0)\,x^n,$$

we generate our previous approximations, as is easily verified.

Example 1. Apply Taylor's approximation to the exponential function near $x = 0$ to give a polynomial of degree five.

Let $f(x) = e^x$, then $f'(x) = f''(x) = f'''(x) = f^{(\text{iv})}(x) = f^{(\text{v})}(x) = e^x$; and as $e^0 = 1$, we have from Taylor's approximation,

$$f(x) \approx f(0)+f'(0)\,x+\frac{1}{2!}f''(0)\,x^2+\ldots+\frac{1}{5!}f^{(\text{v})}(0)\,x^5,$$

$$e^x = 1+x+\frac{x^2}{2!}+\frac{x^3}{3!}+\frac{x^4}{4!}+\frac{x^5}{5!}.$$

The closeness of the approximation to e^x over the interval $-1 \leqslant x \leqslant 1$ is shown in the following table where their values are tabulated for values of x from -1 to 1 by steps of 0·2 (more shortly: $-1(0{\cdot}2)+1$).

x	e^x	$1+x+\ldots+\dfrac{x^5}{5!}$	x	e^x	$1+x+\ldots+\dfrac{x^5}{5!}$
0	1·0000	1·0000	0	1·0000	1·0000
0·2	1·2214	1·2214	−0·2	0·8187	0·8187
0·4	1·4918	1·4918	−0·4	0·6703	0·6703
0·6	1·8221	1·8220	−0·6	0·5488	0·5488
0·8	2·2255	2·2251	−0·8	0·4493	0·4490
1	2·7183	2·7167	−1	0·3679	0·3667

Example 2. Find a polynomial approximation of degree five to $\log(1+x)$ near $x = 0$.

If $f(x) = \log x$, by Taylor's approximation

$$\log(1+x) = f(1)+f'(1)\,x+\frac{f''(1)}{2!}\,x^2+\ldots+\frac{f^{(v)}(1)}{5!}\,x^5.$$

Now $$f'(x) = x^{-1};\quad f''(x) = -x^{-2};\quad f'''(x) = 2x^{-3};$$
$$f^{(iv)}(x) = -3!\,x^{-4};\quad f^{(v)}(x) = 4!\,x^{-5}.$$

Hence $$f(1) = 0;\quad f'(1) = 1;\quad f''(1) = -1;\quad f'''(1) = 2;$$
$$f^{(iv)}(1) = -3!;\quad f^{(v)}(1) = 4!.$$

Therefore $$\log(1+x) \approx x-\tfrac{1}{2}x^2+\tfrac{1}{3}x^3-\tfrac{1}{4}x^4+\tfrac{1}{5}x^5.$$

It is worth remarking that an approximation to $\log x$ near $x = 0$ is meaningless because $\log 0$ is undefined or, what comes to the same thing in more suggestive language, $\log 0$ is $-\infty$.

Figure 4 shows how close is the approximation of $x-\frac{1}{2}x^2+\frac{1}{3}x^3-\frac{1}{4}x^4+\frac{1}{5}x^5$ to $\log(1+x)$ over the range $-0{\cdot}6$ to $+0{\cdot}7$ (the maximum error is in fact under 2%). The figure also shows how the successive approximations of degree 2, 3, 4, 5 improve the 'fit' in the sense that they cling closer to the parent curve for longer. Furthermore, it makes us aware of the fact that no approximation generated by Taylor's method will improve the fit for $x > 1$, because for such values of x the term of highest degree in the approximation becomes preponderant and the graphs of successive higher degree approximations swing more and more wildly up and down away from the curve for $\log(1+x)$. For values of x between -1 and $+1$, however, the graphs suggest (what is true) that the more terms we take the better the approximations become. This is a different state of affairs from what we had in approximating to $\sin x$ and e^x. We showed that for $\sin x$ the approximation improves as we take higher degree polynomial approximations for all values of x. This is also true, but was not proved, for e^x; but for $\log(1+x)$ it is only true over the range -1 to $+1$ (we have not proved this either, but Figure 4 certainly suggests its truth).

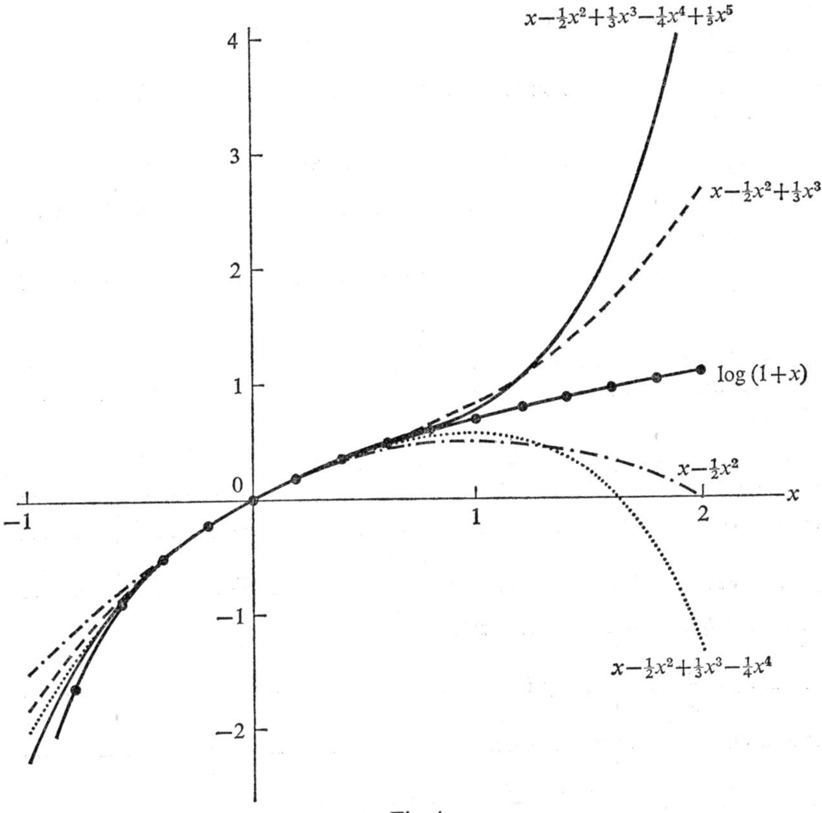

Fig. 4

Polynomial approximations such as $\log(1+x) \approx x-\frac{1}{2}x^2+\ldots+\frac{1}{5}x^5$, and $e^x \approx 1+x+\frac{x^2}{2!}+\ldots+\frac{x^n}{n!}$, are often called expansions in series, or series expansions, of the relevant functions. These expressions are also used for non-terminating series of this sort, but we shall not consider such 'infinite' series here.

Exercise B

1. Find a cubic approximation to $\tan x$ near $x = 0$. Tabulate its values to four figures and the values of $\tan x$ for $x = 0(0{\cdot}1)\,0{\cdot}5$. What is the maximum percentage error in the approximation over the interval $-0{\cdot}5$ to $0{\cdot}5$?

2. (i) By writing $f(x) = 1/x$ and using Taylor's approximation with $a = 1$ and $\epsilon = x$, find an approximation to $1/(1+x)$ near $x = 0$, as far as x^4.

(ii) By multiplying your approximation to $1/(1+x)$ by $1+x$ find its error and show that, if $x > 0$, it is less than x^5.

(iii) Write down a series expansion for $1/(1+x^2)$ as far as x^8.

3. Obtain quartic approximations near $x = 0$ to: (i) $(1+x)^{-2}$; (ii) $(1+x)^5$; (iii) $(1+x)^{\frac{1}{2}}$.

4. Obtain a series expansion to $(1+x)^{-3}$ as far as x^3, and hence find an approximate value to $1/(0{\cdot}98)^3$.

5. With $f(x) = \sin x$, find an expansion for $\sin(x+\frac{1}{4}\pi)$ as far as x^3.

6. Obtain series expansions as far as x^4 for: (i) $\cos x$; (ii) $\sec x$; (iii) x/e^x; (iv) $\log(1/(1+x))$; (v) $\log \sec x$.

7. Obtain a series expansion to $\log\left(\dfrac{1+x}{1-x}\right)$ as far as x^7. Hence, by putting $x = \frac{1}{3}$, find an approximation to $\log 2$ to four figures.

8. Use Taylor's approximation to obtain a polynomial approximation to $e^{-x}\sin x$, as far as x^3. Check your result by multiplying the separate expansions for e^{-x} and $\sin x$.

9. By considering the function f such that $f(0) = 0$ and $f'(x) = e^x$, prove that $e^x > 1$ if $x > 0$, and $e^x < 1$ if $x < 0$. Hence, by a similar argument, prove that $e^x > 1+x$ for all non-zero x. What can you say about $e^x-1-x-\dfrac{x^2}{2!}$? Without proof, make inequality statements about e^x and $1+x+\dfrac{x^2}{2!}+\ldots+\dfrac{x^n}{n!}$ for n even and n odd (check with the table in Section 1.3).

10. If $u_p = \dfrac{x^p}{p!}$ ($x > 0$ and p is a positive integer), show that, whatever the value of x, a value of p (depending on x) can be found such that $\dfrac{u_{p+1}}{u_p} < \frac{1}{2}$. If this value is k, use the fact that $\dfrac{u_{n+k}}{u_k} = \dfrac{u_{n+k}}{u_{n-1+k}}\dfrac{u_{n-1+k}}{u_{n-2+k}}\cdots\dfrac{u_{1+k}}{u_k}$ to prove that $u_{n+k} < (\frac{1}{2})^n u_k$. Say what you deduce from this result about u_{n+k} as n tends to infinity, and hence about u_p as p tends to infinity. Explain the relevance of this to the Taylor expansion for $\sin x$ (see Section 1.2).

11. Use a linear approximation to $\sin x$ near $x = 0$ to obtain an approximation to $x^2 \sin x$ there. Hence sketch the graph of $x^2 \sin x$ near the origin.

12. Use a quadratic approximation to $\cos x$ near $x = 0$ to obtain an approximation to $x(1-\cos x)$ there. Hence sketch the graph of $x(1-\cos x)$ near the origin.

13. Use linear approximations to $\cos x$ and $\sin 2x$ near $x = \frac{1}{2}\pi$ to sketch the graph of $\cos x \sin 2x$ there.

14. (i) Find Taylor's approximation of degree four near $x = 0$ to

$$f(x) = \log\cos x.$$

(ii) Use the approximation $1-\dfrac{x^2}{2}+\dfrac{x^4}{24}$ to $\cos x$ near $x = 0$ and the quadratic approximation near $y = 0$ to $\log(1+y)$ to obtain a quartic approximation to $\log\cos x$ near $x = 0$. Compare this with (i).

15. Use approximations to $\sin y$ and $\log(1+y)$ near $y = 0$ to find approximations to (i) $\dfrac{\sin x^2}{x}$; (ii) $\dfrac{\log(1+x^2)}{x}$, near the origin but excluding the origin itself. Sketch the graphs of these functions of x.

2. MAXIMA, MINIMA AND INFLECTIONS

2.1 Turning points. We have seen that the second derivative $f''(x)$ of $f(x)$ gives the rate of change of the gradient of the graph of f. If f and f' are continuous functions near a point P, so that the graph of f has no breaks or sudden changes of direction, then the graph there will be concave upwards if $f''(x) > 0$, and concave downwards if $f''(x) < 0$, provided f'' exists. For instance, if $f(x) = x^2$, so that $f''(x) = 2$, the graph is everywhere concave upwards; and if $f(x) = -x^2$, so that $f''(x) = -2$, the graph is everywhere concave downwards.

This property of the second derivative sometimes affords a useful criterion of the nature of a turning point. We shall say that $f(p)$ is a *maximum value* of $f(x)$ if, for all x near $x = p$, $f(p) > f(x)$ except at $x = p$; and that $f(p)$ is a *minimum value* of $f(x)$ if, for all x near $x = p$, $f(p) < f(x)$ except at $x = p$.

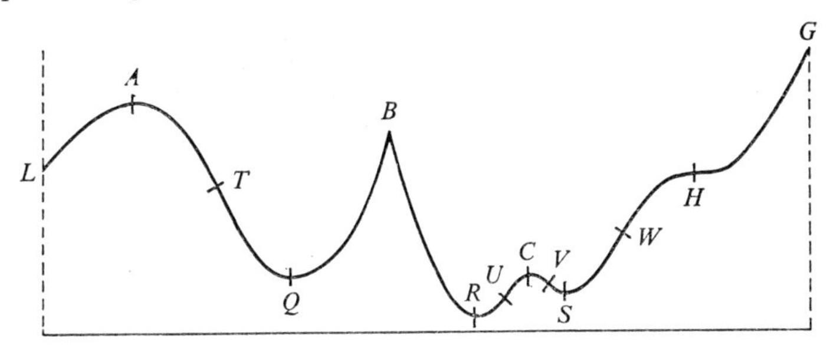

Fig. 5

In Figure 5, maximum values occur at A, B, C; and minimum values at Q, R, S. We shall describe points like A, B, C as maximum points of the graph and those like Q, R, S as minimum points of the graph. B is an 'unusual' type of turning point because the derivative is discontinuous there, and when analysing the nature of a turning point in terms of the second derivative we must ignore such points because the second derivative does not exist there. On this understanding and with the convention that $x = a$ corresponds to the point A, and similarly for the other letters, we have:

$$A \text{ is a maximum point} \Leftrightarrow f'(a) = 0 \text{ and } f''(a) < 0;$$

$$Q \text{ is a minimum point} \Leftrightarrow f'(q) = 0 \text{ and } f''(q) > 0.$$

Let us use this to find where the maximum and minimum values of $f(x) = e^x \sin x$ occur.

$$f'(x) = e^x(\sin x + \cos x) \quad \text{and} \quad f''(x) = 2e^x \cos x.$$

So $f'(x) = 0$ where $\tan x = -1$; that is where $x = n\pi - \pi/4$.

Now $\cos (n\pi - \pi/4)$ is positive or negative according as n is even or odd; and as e^x is always positive, we deduce that the turning points of f occur where $x = n\pi - \pi/4$ and are maximum points if n is odd and minimum points if n is even.

It is important to realize that 'maximum' and 'minimum' are local properties; and in Figure 5 we see that the maximum value at C is in fact less than the minimum value at Q. Furthermore, over a given interval, the greatest value of $f(x)$ may occur not at a maximum point but at an end point of the interval as at G in the figure.

2.2 Inflections. From Figure 5 we can see that the curve is concave downwards between L and T, U and V, and W and H; elsewhere it is concave upwards. The points T, U, V, W, H where the curve changes its concavity are called points of inflection, so that:

T is a point of inflection $\Leftrightarrow f''(t) = 0$ *and* $f''(x)$ has opposite signs on either side of T.

Notice that $f''(t) = 0$ does not by itself imply an inflection, but only if $f''(x)$ changes sign as x passes through the value t. For example, if $f(x) = x^4$, $f''(x) = 12x^2$ and $f''(0) = 0$, but the origin is a minimum point, not an inflection.

At H the inflection is called a stationary inflection because the tangent there has zero gradient. In general a *stationary point* is any point where the gradient is zero.

Example. Examine $f(x) = x(2-x)^3$ for turning points and inflections.

$$f'(x) = (2-x)^3 - 3x(2-x)^2 = (2-x)^2\,(2-x-3x) = 2(2-x)^2\,(1-2x),$$

and

$$f''(x) = -4(2-x)\,(1-2x) - 4(2-x)^2 = -4(2-x)\,(1-2x+2-x)$$
$$= -12(2-x)\,(1-x).$$

Now $f'(x) = 0$ where $x = 2$ or $\frac{1}{2}$; and $f''(2) = 0$ and $f''(\frac{1}{2}) = -9 < 0$. As x increases through the value 2, $f''(x)$ changes from positive to negative ($(2-x)$ changes from positive to negative and, near $x = 2$, $-12(1-x) \approx 12$, which is positive).

Hence at $x = 2$ there is a stationary inflection, and at $x = \frac{1}{2}$ a maximum point. Also, $f''(x) = 0$ where $x = 1$, and as x increases through the value 1, $f''(x)$ changes from negative to positive; so there is an inflection at $x = 1$.

The reader should confirm these findings by sketching the graph of f.

Note. Stationary points can be classified as maximum, minimum or inflections by considering only the signs of $f'(x)$ as x increases through the values where $f'(x) = 0$. This, in many cases, is a better method than one involving the sign of $f''(x)$. For non-stationary inflections it is of course essential to consider $f''(x)$.

Exercise C

1. If $f(x) = x^4 - 8x^3 + 18x^2 - 16x + 5$, find $f'(x)$ and $f''(x)$. Hence sketch the shape of the graph of f near $x = 1, 3, 5$.

2. If $f''(x)$ is (i) $(x-2)\log x$, (ii) $(x-2)^2 \log x$, find, without sketching the graph, the nature of the graph near $x = 2$.

3. If $f(x) = (x^2-1)^2$, find $f'(x)$ and $f''(x)$. Determine the nature of the stationary points by considering (i) only the sign of $f''(x)$, and (ii) only the sign of $f'(x)$ on either side of each stationary point.

Find also the values of x where there are inflections. Sketch the graph of f.

4. If $f'(x) = (x+1)^3 (3x-2)^2$, show that $f''(x) = 15x(x+1)^2 (3x-2)$. For what values of x is $f'(x)$ zero? Use the values of $f''(x)$ at and near these values of x to find the nature of the graph of f there. Obtain the same results by considering the sign of $f'(x)$ only. Sketch the shape of the graph of f at the non-stationary inflection.

5. By considering Taylor's approximation, show that if $f''(p) = 0$ but $f'''(p) \neq 0$, then there is an inflection where $x = p$.

If instead $f'(p) = f''(p) = f'''(p) = 0$ but $f^{(iv)}(p) \neq 0$, what can you say about the graph at $x = p$? State necessary and sufficient conditions for (i) a maximum, (ii) a minimum, (iii) a stationary inflection, at $x = p$ in terms of the successive derivatives of f at $x = p$ (assuming they exist).

6. Sketch the graphs of e^x, $\sin x$ and $e^x \sin x$, in the same figure, for $0 \leqslant x \leqslant \pi$. Where does $e^x \sin x$ have its maximum and inflection points?

7. Locate the stationary points of the graph of $f(x) = \log(1+\sin x)$. Find whether these are maximum or minimum points or points of inflection by (i) considering the signs of $f'(x)$ on either side of each stationary point; (ii) by first finding $f''(x)$. Confirm your results by sketching the graph of f.

8. The function f is such that $f'(x) = 3\sin^2 x(\cos x - \sqrt{3}\sin x)\, e^{-\tan^2 x}$. Find the zeros of $f'(x)$ and the nature of the corresponding points of the graph of f (preferably from the sign of $f'(x)$ on either side of each point).

9. If $f(x) = \sin x \cos^2 x$, find $f'(x)$ and $f''(x)$. For what values of x does $f(x)$ have turning values? Use $f''(x)$ to find the values of x which correspond to (i) a maximum value, (ii) a minimum value, of $f(x)$, and verify that maximum and minimum values occur alternately.

10. Show that inflections of $f(x) = x \sin x$ occur where $\tan x = 2/x$, and (by referring to Chapter 9, p. 183, Figure 18) that these values are close to $n\pi$ for large integral values of n. Use Newton's method to show that for large n a closer approximation to the positions of the inflections is where $x = n\pi + (2/n\pi)$.

ANSWERS

CHAPTER 1

Exercise A

1. (*a*) $\begin{Bmatrix} 2 & 3 \\ 3 & 10 \\ 4 & 18 \\ 5 & 42 \\ 6 & 75 \end{Bmatrix}$, {2, 3, 4, 5, 6}, {3, 10, 18, 42, 75}.

(*b*) $\begin{Bmatrix} 2 & 210000 \\ 3 & 340000 \\ 5 & 340000 \\ 6 & 133000 \end{Bmatrix}$, {2, 3, 5, 6}, {210000, 340000, 133000}.

(Different functions are possible: for example, the unit of distance could have been 10000 km.)

(*c*) $\begin{Bmatrix} 1 & 14 \\ 2 & 8{\cdot}50 \\ 3 & 21 \\ 4 & 17{\cdot}75 \end{Bmatrix}$, {1, 2, 3, 4}, {14, 8·50, 21, 17·75}.

(Different functions are possible: for example, the coding of the four people could have been different.)

2. (i) 7; (ii) 4; (iii) 13.

3. (i) 5; (ii) 3, 6; (iii) 10; (iv) {5, 2, 10}.

4. {0, 1, 4}. 2, −2. 1.

5. E.g. $\begin{Bmatrix} 2 & -7 \\ 3 & -7 \\ -5 & 17 \end{Bmatrix}$, $\begin{Bmatrix} 2 & 17 \\ 3 & 17 \\ -5 & -7 \end{Bmatrix}$, $\begin{Bmatrix} 2 & 17 \\ 3 & -7 \\ -5 & 17 \end{Bmatrix}$. 6.

6. (i)–(v) {1, 3, 6}, {2, −1}. 3.

Exercise B

1. (i) 0; (ii) 4; (iii) −4; (iv) no meaning.

2. (i) 30; (ii) 4; (iii) 10; (iv) 7; (v) 11. {4, 7, 11}. 30, 4. $a = 0$ or -5.

3. 2. 5.

4. {0, 1, 4}. 1. ±2.

5. (i) $f(2) = 7$; (ii) $f(4) = 11$; $2x+3$.

6. $1/(x+1)$. 1/501. 99. No (0 does not belong to the domain). No.

7. (i) If x is replaced by a sequence of numbers each greater than its predecessor, we get a sequence of images each less than its predecessor; (ii) the image of 0 is 5; (iii) if x is replaced by a sequence of numbers each twice its predecessor, we get a sequence of images each thrice its predecessor.

8. (i) There is a function f such that at time t seconds the temperature is $f(t)$ °C. Or: at any time there is a definite temperature.

(ii) Distance is $4t$ km when the time is t hours.

9. $2\pi r^2 + \dfrac{20}{r}$ cm^2.

Exercise C

1. $f(x) = 5000/x$. It could be continuous, but may be discrete. (All numbers are positive). $f(x)$ closes towards zero as x increases indefinitely, and $f(x)$ increases indefinitely as x closes towards zero. It takes a short time at high speeds and a long time at slow speeds; for a sufficiently high speed the time can be as short as we like and for a sufficiently slow speed the time can be as long as we like.

2. r. Discrete. As r increases from 1 to 9, $f(r)$ increases to a maximum when $r = 5$ and then decreases. The function is symmetrical about $r = 5$. There are few small or large sets of cabinet ministers but many of the medium sized ones.

3. ...the number of sixes. The greatest probability is for one six although no sixes is almost as probable; as the number of sixes increases beyond one, the probability decreases and becomes very small for six sixes.

4. In cm and second units, f is a straight line rising from 4 to 14 as r goes from 0 to 5. The graph of g starts at 4 when $r = \sqrt{(4/\pi)}$ and becomes steeper as r increases. The graph of h starts at $\sqrt{(4/\pi)}$ when $t = 0$ and becomes less steep as t increases.

The area increases steadily with time. If the radius increases steadily the area increases more and more rapidly. The radius increases progressively more slowly with time.

Exercise D

1. Another possible variable is the height.

2. Temperature is a function of time.

3. The greatest height is not a function of the range, because for any range (in general) there corresponds two greatest heights (for the two possible paths).

6. Area is very large when the cross-section is either very small or very large. It falls and then rises in between.

7. Volume as a function of the base-radius, for example, rises from zero when $r = 0$ and falls again to zero when r has its greatest value (when the tent is flat on the ground).

8. The probability is small when there are two people and as the number increases the graph rises. With large numbers the rate of rise decreases and the graph steadies out towards a height of 1 as the number increases indefinitely. If exactly two people are to have the same anniversary, the graph rises at first and later falls, flattening out towards the domain axis.

9. (i) The graph rises from near zero to a maximum, starting slowly, and then flattens out towards zero. (ii) Similar graph but flattening to zero when the angle is 180°.

10. Starts from zero, rises to a maximum and then falls, flattening out towards zero as time increases indefinitely. Starts at height of 1 and gradually flattens out towards zero. Starts more slowly than the first graph and reaches its maximum later; it then flattens out towards zero.

CHAPTER 2

Exercise A

1. (0, 4), (8, −4); (4, 0).

2. $f(0) = -2, f(6) = 2$. Steady increase of $\frac{2}{3}$.

3. Through (0, −3) gradient 2, ($\frac{3}{2}$, 0); through (0, 5) gradient −1, (5, 0); through (0, −7) gradient $-\frac{3}{4}$, ($-\frac{28}{3}$, 0).

4. 4, ($-\frac{1}{2}$, 0); −1, (−3, 0); 2, (0, 0).

5. $f(x) = c$, where c is any number. No. The image of the point where it crosses the x-axis is not unique.

6. Lines through (0, 3) with gradients: −2, −1·5, −1·8, −1·9.

7. $a = 4$, $b = 3$. $f(x)$ increases in steps of b.

8. −2, 13, 17, 3.

9. 6.

10. Distance decreases steadily. −4. No.

11. $d = 22-6t$. 16 km east; 8 km west. 6 km/h.

12. It is linear. 22 cm. −8. Under compression from force equal to the weight of 8 g.

13. Yes. $T = 12$, $x = 36$. Lines cross at (36, 12).

14. 2π. 2π. (i) 2π m; (ii) about 44 m.

15. $D = 6T$. He is walking in the opposite direction (or backwards). By a line with negative gradient.

16. 0·24, 0·1 cm.

17. 0·75.

Exercises B, C

Nil.

Exercise D

The higher power has the smaller value when x is smaller than the following; and the larger value when x is larger than the following:

1. 1; **2.** 1; **3.** 8; **4.** 250; **5.** $\frac{1}{3}$; **6.** $1/\sqrt{2}$; **7.** 1.

Exercise E

Nil.

Exercise F

3. (1, 3); (0, 1); (0, 1).

Exercise G

8. x^{-1}, x^{-2}, x^{-3} are decreasing/increasing values for x larger/smaller than 1.

Exercise H

4. For example:

(*a*) $x(x-1)(x+2)$; (*b*) $x^2(x+3)$; (*c*) $-x^3(x+2)$;

(*d*) $(2-x)^3(x+1)$; (*e*) $-x(x+2)^2(x-3)^2$.

6. Positive (or zero) except for: $2 < x < 3$; $x < -1$; $-3 < x < 3$; $x < 0$ and $2 < x < 3$; $x < -2$ and $0 < x < 1$.

Miscellaneous

1. $a = 2$, $b = 60000$. About £693, for about 173 cars each day. No (from a sketch-graph).

2. $v = \dfrac{2u}{u+2}$. More than 4 m or less than $1\frac{1}{3}$ m to the left of the lens. Less than 2 m to the right of the lens.

3. $\dfrac{200}{x+5} + \dfrac{200}{10-x}$ seconds, when the water runs at x m/s. Least time $53\frac{1}{3}$ s when $x = 2\frac{1}{2}$, by symmetry of the two graphs added.

4. $\dfrac{(x-40)(x+40)}{x}$ km/h, where his air-speed is x km/h. 1600 km/h.

5. 8 cm^3.

CHAPTER 3

Exercise A

1. $|x| < 0{\cdot}87$; $|x| < 0{\cdot}44$; $|x| < 0{\cdot}31$.

2. All but $9000x$, $x/9000$, x^{-1}, $x^2 - 10^{-6}$.

4. $\sqrt{x}$, no; $x\sqrt{x}$, yes.

5. Yes. All $|k| \leqslant 0{\cdot}02$. No.

6. Yes. Yes.

10. $xf(x)$, no; $x^2f(x)$ and $x^3f(x)$, yes. x^{-1}. $f(x)$ bounded near $x = 0$.

11. Yes. $g(0) = 0$ and $g(x)$ bounded near $x = 0$.

Exercise B

6. $(a^2-6a+7)+(2a-6)\,\epsilon+s_\epsilon$.

Exercise C

1. $-14-12\epsilon$, $-8+4\epsilon$, $-20-68\epsilon$, 29/32.

2. $|\epsilon| \leqslant 0{\cdot}2$. 9·8 $|\epsilon| < 0{\cdot}005$. 7·94. Correct to 0·005. $|\epsilon| < 0{\cdot}002$.

Exercise D

4. Those which map x onto 5, -3, 0. They are of the form $f(x) = 5x+\text{constant}$, $-2x+\text{constant}$, constant.

5. $f'(a) = 2a-3$.

6. $f'(a) = 2a$.

7. (*a*) $f'(a) = 3a^2$; (*b*) $f'(a) = 4a^3$.

8. (*a*) $f'(a) = 3a^2-6a+5$; (*b*) $f'(a) = 6a^2+4a^3$.

9. Derived functions of x are $3x^2+2$, $3x^2-2$.

Exercise E

1. 7, $6x$, 5, $2-3x^2$, $7x^6$, $-1+9x^2-42x^5$.

2. 0 and -28; 7 and 5; 27.

3. 92, -22. $x = 0$.

4. $x < \frac{1}{3}$.

Exercise F

1. $6x+21x^2+2x^{-2}$; $1-x^{-2}$; $7x+\frac{5}{2}x^{-2}$.

2. $-34{\cdot}5$, $16{\cdot}125$.

CHAPTER 4

Exercise A

1. $-2{\cdot}1$.

2. 4·5, 7·2.

3. 3·16.

4. 1·23; $-2{\cdot}38$ and 2·05.

5. 2 negative, 1 positive; 2 negative, 2 positive; 1 negative, 1 positive; 1 negative, 2 positive.

6. Between 1·8 and 1·9.

Exercise B

1. 2, 1 or 0 roots according as $c \gtreqless -16$.

2. 2, 1 or 0 roots according as $b^2 \gtreqless 144$.

4. 3, 2 or 1 roots according as $k \gtreqless 1$.

5. 2, 1 or 0 roots according as $k^2 \gtreqless 4$.

6. (*a*) 2, 1 or 0 roots according as $b^2 \gtreqless 4ac$;
(*b*) (i) and (iv) positive; (ii) and (iii) negative.

Exercise C

1. $2x/(1+x)$.

2. (i) all sides 5 cm. (ii) all sides 6 cm.

3. Domain, $0 \leqslant x \leqslant 100$; range, $0 \leqslant f(x) \leqslant 250$.

4. 50 cm^2. **5.** 4·24 cm^2. **6.** 2/27 m^3.

7. $4N/3(3N-1)$. **8.** 0·35. **9.** $K > 1$ and $K < \frac{1}{2}$.

Exercise D

1. 3 cm × 5 cm × 10 cm. 3 cm × 50 cm × 1 cm, with 2 bands round the largest section.

2. Side of base $= \sqrt[3]{2V}$, where V is the volume.

3. 864 cm^3.

4. Domain, $-1 < x < 1$; range $0 < f(x) < \frac{32}{81}\pi$.

5. 2·17 seconds. **6.** No. **10.** Up.

11. Increase; $0 < r < 4$. **12.** With sides of 9 and 39 hurdles.

13. $p = \frac{1}{2}$. **14.** $0{\cdot}8 < R < \frac{4}{3}$.

CHAPTER 5

Exercise A

1. N/m^3. **2.** The number 1.

3. $-$£50/child *or* $-$50(£/child). **4.** 16π cm^2. 0·5 cm^3.

5. 0·4 cm fall; 0·4 cm rise. **6.** 150 cm^2. 15 cm^3.

7. $A = \pi r^2 = \pi(3t+1)^2$. 3 cm/s. 8π cm. 24π cm^2/s. $0{\cdot}4\pi$ cm^2.

8. 48π cm^2. $0{\cdot}48\pi$ cm^3. 9π cm^2. $0{\cdot}18\pi$ cm^3. Exact. 0·48/9 (=0·053) cm.

Exercise B

1. (i) 9 m; (ii) 24 m/s; (iii) 12 m/s^2; (iv) 27 m/s.

3. It moves from rest to the left for 4 s, when it comes to rest again 32 cm from O; then it starts to move to the right, passing through O, 2 s later and continues at an ever increasing speed.

4. 32 m/s. After 8 s; -8 m/s^2; $170\frac{2}{3}$ m.

5. All zero. Yes.

6. Zero velocity corresponds to extremes of distance. Maximum and minimum velocity correspond to the inflections of the distance-time graph.

Exercise C

1. x^2+1. 2. $2x^5-x^3+4$. 3. $x+\frac{2}{x}-3$.

4. (i) 1 m/s; (ii) $1\frac{1}{3}$ m/s; (iii) $\frac{1}{3}$ m/s². $s(t) = t+\frac{1}{36}t^3$. $2\frac{2}{9}$ m. $1\frac{1}{9}$ m/s.

5. (i) 78 m/s, 30 m/s²; (ii) 16 m/s, 16 m.

6. 19·2 m.

7. (i) 102 cm/s; (ii) when $t = 3$; (iii) 1218 cm.

8. (i) 112 m²; (ii) 2784 m². 9. 1·7 hour.

10. 400. 11. 50. 12. 11:1.

13. $-\frac{1+10x^2}{5x^2}$. 6 min 9 s.

Exercise D

1. 27 cm². 2. 66 700 m³. 3. £1833 000.

4. £1200. 5. $\frac{1}{2}kab^2$ N. 6. 0·198 m³.

7. 0·8 m. 8. 2560 kg (2559). 9. 7·98 kg.

10. 7·973g Nm; 7·973g N.

Exercise E

1. 780. £58·80. 2. 503. 1 m. 3. $\frac{2}{3}\mu Wa$.

4. $(10pa^3+ka^5)/30$. 5. Proportional to X.

CHAPTER 6

Exercise A

1. (i) $-\frac{1}{2}$; (ii) 2; (iii) $\frac{1}{2}$; (iv) 3.

2. (i) 3; (ii) 1; (iii) $2x^2+1$; (iv) $(2x+1)^2$.

3. $f(x) = x^{\frac{1}{2}}$, $g(x) = x^2+1$. 4. $f = P_3$, $g(x) = 3x+7$.

5. (i) P_2g where $g(x) = 3x^2-1$; (ii) $P_{-1}g$ where $g(x) = 1-5x$; (iii) $P_{-1}P_3\,g$ or $P_{-3}\,g$ where $g(x) = 1-x^2$; (iv) $P_3P_{\frac{1}{2}}g$ where $g(x) = 2x+1$.

6. Domain of f, $x > 9$; of g, real x; of fg, $|x| > 9$; of gf, $x > 9$.

7. $fg(x) = x^2-2$; $gh(x) = (x+2)^2+1$.

8. P_6, P_6, P_{-3}, P_{-6}.

Exercise B

1. (i) 7; (ii) 4·75; (iii) 3·31; (iv) 3·0301; (v) 3·003001.

2. 12, 6, 72; $2(x^3-5)$, $3x^2$, $6x^2(x^3-5)$.

3. $75/8$; -2.

4. (i) $6(3x+1)$; (ii) $5(2x+3)(x^2+3x-1)^4$; (iii) $-8(5-2x)^3$; (iv) $-8x/(x^2+7)^5$; (v) $2/(1-x)^3$.

6. $-4x^{-5}$.

7. (i) $-6(1-2x)$; (ii) $6x(x^2+1)^2$; (iii) $2(2-3x^2)(2x-x^3)$.

8. (i) $8(2x-3)^3$; (ii) $5(6x-3)(3x^2-3x+2)^4$; (iii) $-6(1-2x)^2$; (iv) $-2x/(x^2+1)^2$; (v) $-5x^{-6}$; (vi) $-3x^{-4}+1$.

9. Minimum where $x = -2$.

10. Minimum where $x = \frac{1}{4}$.

11. $4{\cdot}32\ \text{m}^2$.

12. $x = 3$. $69\frac{1}{2}°$.

13. $P_{-1}P_3\,g$ or $P_3P_{-1}\,g$ where $g(x) = x^2+2$; $-6x(x^2+3)^4$.

Exercise C

1. (i) $\frac{1}{2}x^{-\frac{1}{2}}$; (ii) $-\frac{1}{2}x^{-\frac{3}{2}}$; (iii) $\frac{1}{3}x^{-\frac{2}{3}}$; (iv) $3/2\sqrt{(3x-7)}$; (v) $1/2\sqrt{(3-x)^3}$.

2. (i) $2x^{-\frac{1}{3}}$; (ii) $\frac{7}{3}x^{-\frac{2}{3}}$; (iii) $20x/\sqrt{(4x^2-3)}$; (iv) $9/\sqrt{(2-3x)^3}$.

3. $f'(0) = 0$. Domain, $x \geqslant 0$.

4. Minimum at $x = 1$. $f'(4) = \frac{3}{16}$; $f'(9) = \frac{4}{27}$.

5. $f'(x) = \frac{1}{2}(1-x)^{-\frac{3}{2}}$.

6. $3+\frac{1}{6}\epsilon$.

7. $(10-\epsilon)/100$.

8. $3{\cdot}05$, $0{\cdot}2008$.

10. 72 seconds.

Exercise D

1. (i) $6x^2+6x+2$; (ii) $2x-5x^4$; (iii) $1-x^{-2}$; (iv) $7x^6$.

2. (i) $2x(2x-1)^2(5x-1)$; (ii) $(1-x)(1+x)^2(1-5x)$; (iii) $(4x^2+2x-1)^2(28x^2+8x-1)$.

3. (i) $(5x^2+4x)/2\sqrt{(x+1)}$; (ii) $(x+2)/2\sqrt{(x+1)^3}$; (iii) $(2x-3)/2\sqrt{(x^2-3x-1)}$; (iv) $(-4x-1)/\sqrt{(x+1)}(1-2x)$; (v) $(3x^2+4x+3)/2\sqrt{(x+1)^3}$.

4. Maximum at $x = \frac{2}{3}$; minimum at $x = 0$; $x < 1$.

5. $x = \pm\sqrt{(\frac{3}{2})}$.

6. 3 roots.

9. $\sqrt{2}$.

10. $0 < w < 2/\sqrt{3}$; $w = 1/\sqrt{3}$.

11. $2/\sqrt{5}$ m/s.

12. $v^2t/\sqrt{(v^2t^2-h^2)}$; $-v^2h^2/\sqrt{(v^2t^2-h^2)^3}$.

13. $\sqrt{(5+4\cos x°)}$.

14. $c/\sqrt{(1-c^2)}-dc$, where $c = \cos x°$.

16. Walk $200-40\sqrt{5}$ m down one side, then across.

Exercise E

1. (i) $(1-2x-x^2)/(1+x^2)^2$; (ii) $-2(x^2+1)/(x^2-1)^2$; (iii) $-2(x-1)/(1-2x)^3$.

2. (i) $(x^3+2x)/\sqrt{(1+x^2)^3}$; (ii) $(-x^2+x-1)/\sqrt{[(x^2-1)^3(2x-1)]}$;
(iii) $(4x-4)/(1+2x)^2\sqrt{(1-4x)}$.

4. (i) $(f'g).g'.h+(fg).h'$; (ii) $f'(g.h).(g'.h+g.h')$.

CHAPTER 7

Exercise A

1. $2{\cdot}2\times 9^x$; $-1{\cdot}6\times(\frac{1}{5})^x$; $2{\cdot}8\times 16^x$; $-3{\cdot}3\times(\frac{1}{27})^x$; $2{\cdot}3\times 10^x$; $-0{\cdot}5\times(\frac{3}{5})^x$; $4{\cdot}1\times 60^x$.

2. 2^{2x}, $1{\cdot}4\times 4^x$; 3^{-x}, $-1{\cdot}1\times(\frac{1}{3})^x$; 3^{3x}, $3{\cdot}3\times 27^x$; 2^{-4x}, $-2{\cdot}8\times(\frac{1}{16})^x$; $2^{\frac{1}{2}x}$, $0{\cdot}3\times 2^{\frac{1}{2}x}$; 2^{-4x}.

4. $2{\cdot}2\times 9^5$; $2{\cdot}2\times 9^{-5}$; $-1{\cdot}4\times(\frac{1}{4})^5$, $-1{\cdot}4\times 4^5$; $0{\cdot}3\times(\frac{4}{3})^5$, $0{\cdot}3\times(\frac{4}{3})^{-5}$.

5. $x2^x(2+0{\cdot}7x)$; $x^{-2}6^x(1{\cdot}8x-1)$; $5^{-x}(1-1{\cdot}6x)$; $1{\cdot}4x2^{(x^2)}$; $-1{\cdot}6x^{-2}\,6^{1/x}$; $(0{\cdot}7)^2\,2^{(2^x+x)}$; $1{\cdot}2\times 12^{x/2}$.

7. None.

8. If $K = k_3 3^{1/k_3}$, 2 roots if $a > K$; 1 root if $a = K$ or if $a < 0$; no roots if $0 \leqslant a < K$.

Exercise B

1. (i) $P(\frac{3}{2})^t$; (ii) $m(0{\cdot}99)^t$; (iii) $V(\frac{4}{5})^{t/8}$; (iv) $V(\frac{4}{5})^{T/96}$; (v) $N(\frac{4}{5})^t$ and $N(\frac{4}{5})^{60T}$; (vi) $C(\frac{7}{6})^{t/240}$.

2. (i) 0·32; (ii) 4·49 weeks; (iii) 39·2 hours/week.

3. About 2300 years ago.

4. 0·009/year/unit in population; 0·203.

5. 0·476. **6.** 1/105. **7.** $0{\cdot}00175t$. **8.** 0·001.

Exercise C

1. 0·069/(number of days in the month).

2. $N(\frac{3}{17})(\frac{17}{20})^N$; 6. **3.** 0·718 hours. **4.** 0·47 hours.

Exercise D

1. $5e^{5x}$, $2e^{2x}$, $-e^{-x}$, $\frac{1}{2}e^{\frac{1}{2}x}$, $-3e^{-3x}$, $-12e^{-12x}$, $-4e^{-4x}$, $\log_e 2\times 2^x$, $\log_e 0{\cdot}1\times 0{\cdot}1^x$, $\log_e 10\times 10^x$, $\frac{1}{2}\log_e\frac{3}{4}\times(\frac{3}{4})^{x/2}$.

3. $2xe^{(x^2)}$.

4. $-x^{-2}e^{1/x}$, $-2e^{1-2x}$, $-2e^{-x}(1+e^{-x})$, $2e^{2x}/(1-2^{2x})^2$, $e^xe^{(e^x)}$.

Exercise E

5. (i) $k > e$; (ii) $0 < k < e$. **6.** 2.

7. 2; if $K = 8e^3/27$, 2 roots if $a > K$; 1 root if $a = K$ or if $a < 0$; no roots if $0 \leqslant a < K$.

8. $(\log_e 2)/10$.

9. $1/p$; the same; $2/p$; $3/p$; p events per unit time.
(*a*) 0·190; 0·315.
(*b*) 3 cars; 0·226; 0·174; 0·221.
(*c*) 1600 with 1 spot; about 29000 altogether.

CHAPTER 8

Exercise A

1. $1/x$, $1/x$, $3/x$, $-1/x$, $-2/x$.

2. $5(\log x)^4/x$, $5/x$, $x+2x\log x$, $1+\log 4x$, $\dfrac{1}{2x}$, $-2/(1-2x)$, 1, $2x/(x^2-1)$,

$-(\log x)^{-2}/x$, $(\log x)^{-1}-(\log x)^{-2}$, $x^{-1}+2(1-2x)^{-1}$, $\dfrac{1}{x}\,e^{\log x}(=1)$.

5. $2^{a+x} = 2^a 2^x$; 2^x is differentiable at $x = 0$.

6. The derivative of $\log_a x$ at $x = 1$.

Exercise B

3. Gradients are: ∞, ∞, 0. **5.** 0. **6.** 2. 1·9 *or* 4·5.

7. $a < -\dfrac{1}{2e}$: none; $a = -\dfrac{1}{2e}$: 1; $-\dfrac{1}{2e} < a < 0$:2; $a \geqslant 0$:1.

8. $1+k\epsilon$, ϵ.

10. $x^{x^2+1}(1+2\log x)$, $2x^{\log x-1}.\log x$,
$\frac{3}{2}(2-5x+2x^2-3x^3)\,x^2(1-3x)^{-\frac{3}{2}}\,(1+3x^2)^{-2}$,
$(\log x)^x\,[\log(\log x)+(\log x)^{-1}]$.

Exercise C

1. $\frac{1}{5}\log|x|+C$, $-\log|x|+C$, $\frac{1}{3}\log|3x-2|+C$, $-\frac{1}{3}\log|2-3x|+C$,
$\frac{1}{2}\log(1+x^2)+C$, $-\frac{1}{2}\log|1-2e^x|$.

2. $f(x) = \frac{1}{3}\log\left(\dfrac{1+3x}{4}\right)+5$ for $x > -\frac{1}{3}$; $f(x) = \frac{1}{3}\log\left(\dfrac{1+3x}{-5}\right)+7$ for $x < -\frac{1}{3}$.

3. $f(x) = -\log\left(\dfrac{3-x}{3}\right)$ for $x < 3$; $f(x) = -\log\left(\dfrac{x-3}{2}\right)-3$ for $x > 3$.

4. 374 J.

Exercise D

1. 2·46. 0·32 week.
2. 1·1 min.
3. $4{\cdot}605/k$.
4. $\frac{1}{k}\log(1/\lambda)$.
5. 0·575.
6. $\frac{1}{k}\log 10$.
7. $\frac{V}{ak}\log(10/9)$, where a is the constant involved in the diffusion law.
8. $CR\log 2$.

CHAPTER 9

Exercise A

1. $\pi, \frac{1}{2}\pi, \frac{3}{2}\pi, \frac{1}{3}\pi, \frac{2}{3}\pi, \frac{1}{6}\pi, \frac{5}{6}\pi, \frac{11}{6}\pi$.
2. 45°, 120°, 30°, 135°, 144°.
3. (i) 10 cm, 20 cm, $\frac{10}{3}\pi$ cm; (ii) 50 cm², 25 cm², 20π cm².
4. (i) $20\sin(\frac{1}{20}\pi):\pi$; (ii) $200\sin(\frac{1}{200}\pi):\pi$.
5. 0·0525, 0·0175, 0·0350, 0·0175, 0·0175, 0·0525, 0·0087.
6. $3\cos 3x$, $-\frac{1}{4}\sin\frac{1}{4}x$, $-2\sin 2x$, $\frac{3}{2}\cos\frac{1}{2}x$, $\cos(x+\pi)$, $-2\sin(2x-\frac{1}{2}\pi)$, $-20\cos(\frac{2}{3}\pi-4x)$, $9\sin 3x$, $\frac{9}{4}\cos(\frac{3}{4}x+\frac{1}{2}\pi)$.
8. 0·52.
9. 0·41 radians.

Exercise B

1. $-\operatorname{cosec} x\cot x$, $2\sin x\cos x$, $2\sec^2 2x$, $2\sec^2 x\tan x$, $-6\cos^2 2x\sin 2x$, $-\operatorname{cosec}^2 x$, $\cos^2 x-\sin^2 x$, $2x\tan 3x+3x^2\sec^2 3x$, $\cot x$, $e^x\cos(e^x)$, $2\tan 2x$.
2. $2\sin x\cos x$, $-2\sin x\cos x$; that $\sin^2 x+\cos^2 x = 1$.
9. 1.
10. 2.
11. 3.
12. 2 units area.
13. 2π units volume.
14. No roots if $k > \sin 2\alpha_1$, or $k < 0$; 1 root if $k = \sin 2\alpha_1$, or if $k = 0$; 2 roots if $\sin 2\alpha_1 > k > \sin 2\alpha_2$; 3 roots if $k = \sin 2\alpha_2$; 4 roots if $0 < k < \sin 2\alpha_2$.

Exercise C

1. $\sqrt{2}\sin(x+\frac{1}{4}\pi)$.
2. $13\cos(x-\alpha)$, where $\alpha = \tan^{-1}(\frac{12}{5})$.
3. $\sqrt{58}$ for $x = 2r\pi-\tan^{-1}\frac{3}{7}$.
4. $r = \sqrt{34}$, $\alpha = \tan^{-1}(\frac{3}{5})$.
6. $\sqrt{19}\sin(x+0{\cdot}408)$, $\sqrt{19}\cos(x-1{\cdot}162)$.
7. $7{\cdot}59\sin(x+0{\cdot}490)$.

Exercise D

1. Maxima where $x = \pi+\alpha$, $2\pi-\alpha$ ($\alpha = \tan^{-1}\frac{1}{4}$); minima where $x = \frac{1}{2}\pi, \frac{3}{2}\pi$.
2. Maxima where $x = \frac{7}{6}\pi$; minimum where $x = \frac{11}{6}\pi$; point symmetry about $((2r+1)\frac{1}{2}\pi, 0)$.

4. (i) Maximum where $x = \cos^{-1} 0{\cdot}46$; minimum where $x = 2\pi - \cos^{-1} 0{\cdot}46$.
(ii) Maxima, where $x = \sin^{-1} 0{\cdot}92$, $\pi + \sin^{-1} 0{\cdot}54$; minima where

$$x = \pi - \sin^{-1} 0{\cdot}92,\ 2\pi - \sin^{-1} 0{\cdot}54.$$

(iii) Maxima where $x = \frac{1}{5}\pi, \frac{7}{5}\pi$; minima where $x = \frac{3}{5}\pi, \frac{9}{5}\pi$.

10. Maxima, $a-b$, $\frac{2}{3}\sqrt{(b^3/3a)}$; minima, $b-a$, $-\frac{2}{3}\sqrt{(b^3/3a)}$; $a = 4$, $b = 3$.

Exercise E

1. $f(x) = 2\pi r^3 \sin^2 x \cos x$.

2. $\frac{1}{6}\pi$.

3. $27\sqrt{3}\pi/2$ m^2.

4. Yes.

5. 200/3 units downstream.

6. $\cos^{-1}(\frac{3}{5})$; yes.

7. $32R^3/81$.

8. She does.

9. If $k > \frac{1}{2}\pi$, best to run all the way; if $k < \frac{1}{2}\pi$, best to wade all way.

10. 300 m.

11. $(24\sqrt{5})/11$ m/s.

12. If $a \leqslant d$, a maximum where $x = 0$; if $a > d$, maxima where $\tan x = ax/d$ but $x \neq 0$.

13. Sin $x - d \tan x$. $x^{\frac{2}{3}} + y^{\frac{2}{3}} = 1$.

14. If $0 < x < \tan^{-1} \frac{8}{25}$, $f(x) = 25 \sec x$; if $\tan^{-1} \frac{8}{25} < x < \tan^{-1} 2$,

$$f(x) = \sqrt{\left(\frac{200}{\sin x \cos x}\right)};\ \text{if } \tan^{-1} 2 < x < \frac{\pi}{2},\ f(x) = 20 \operatorname{cosec} x.$$

15. After t seconds, $v = 4\pi^2 at$.

CHAPTER 10

Exercise A

2. (i) $-3-x$, $-3-x-x^2$; (ii) $2x$, $2x+3x^3$; (iii) 5, $5+2x^2$.

3. (i) $|x| < 0{\cdot}02$ $(0{\cdot}0168)$, (ii) $|x| < 0{\cdot}05$ $(0{\cdot}047)$. $x > -0{\cdot}5$.

5. $-6+\frac{1}{2}\epsilon$, $-6+\frac{1}{2}\epsilon-2\epsilon^2$.

6. 2·3.

7. $-\frac{1}{3}+2x$.

8. $-\frac{1}{3}+2x$. (If starting from $a = -\frac{1}{3}$, $b = 2$, we increase a, b by p, q, then it can be shown that $3a^2+5b^2+6ab-10a-18b+17$ is increased by $3p^2+6pq+5q^2$, that is by $(3p+q)^2+4q^2$, which is always positive. Hence for all values of p, q, the sum is increased.)

Exercise B

1. $\tan x \approx x+\frac{1}{3}x^3$. Corresponding values of $\tan x$ and $x+\frac{1}{3}x^3$ for 0, 0·1, 0·2, 0·3, 0·4, 0·5 are (0, 0), (0·1003, 0·1003), (0·2027), 0·2027) (0·3093, 0·3090), (0·4228, 0·4213), (0·5463, 0·5417). Maximum error 0·8 %.

2. (i) $1-x+x^2-x^3+x^4$; (ii) Error $x^5/(1+x)$; (iii) $1-x^2+x^4-x^6+x^8$.

3. (i) $1-2x+3x^2-4x^3+5x^4$; (ii) $1+5x+10x^2+10x^3+5x^4$;
(iii) $1+\frac{1}{2}x-\frac{1}{8}x^2+\frac{1}{16}x^3-\frac{5}{128}x^4$.

4. $1-3x+6x^2-10x^3$. 1·06248. **5.** $2^{-\frac{1}{2}}\left(1+x-\frac{x^2}{2!}-\frac{x^3}{3!}\right)$.

6. (i) $1-\frac{x^2}{2!}+\frac{x^4}{4!}$; (ii) $1+\frac{x^2}{2!}+\frac{5x^4}{4!}$; (iii) $x-x^2+\frac{x^3}{2!}-\frac{x^4}{3!}$;

(iv) $-x+\frac{x^2}{2}-\frac{x^3}{3}+\frac{x^4}{4}$; (v) $\frac{1}{2}x^2+\frac{1}{12}x^4$.

7. $2\left(x+\frac{x^3}{3}+\frac{x^5}{5}+\frac{x^7}{7}\right)$. 0·6931.

8. $x-x^2+\frac{1}{3}x^3$.

9. $e^x > 1+x+\ldots+\frac{x^n}{n!}$, for all non-zero x if n is odd and for $x > 0$ if n is even;

$e^x < 1+x+\ldots+\frac{x^n}{n!}$, for all $x < 0$ if n is even.

10. k can be any positive integer $> 2x-1$. $u_{n+k} \to 0$ as $n \to \infty$. $u_p \to 0$ as $p \to \infty$. The error in approximating to $\sin x$ by n terms of the series expansion, of the form derived in Section 1.2, is less than $\left|\frac{x^{2n+1}}{(2n+1)!}\right|$, which tends to zero for all $x > 0$ as $n \to \infty$. Likewise if $x < 0$. This substantiates the statement in the first paragraph of Section 1.2.

13. If $x = \frac{1}{2}\pi+\epsilon$, $\cos x \approx -\epsilon$ and $\sin 2x \approx -2\epsilon$.

14. $-\frac{1}{2}x^2-\frac{1}{12}x^4$.

Exercise C

2. (i) Inflection, (ii) minimum.

3. Stationary points: $x = 0, 1, -1$ are maximum, minimum, minimum. Inflections at $x = \pm 1/\sqrt{3}$.

4. At $x = -1$, minimum; at $x = \frac{2}{3}$, stationary inflection.

5. Maximum or minimum point. In each case $f'(p) = 0$, and if we denote the first higher derivative which is non-zero by $f^{(r)}(p)$, then (i) r is even and $f^{(r)}(p) < 0$, (ii) r is even and $f^{(r)}(p) > 0$, (iii) r is odd.

6. Maximum at $x = \frac{3}{4}\pi$; inflection at $x = \frac{1}{2}\pi$.

7. Stationary points are $x = \frac{1}{2}\pi+2n\pi$, for all integral n. They are all maximum points.

8. Stationary points: $x = n\pi$, inflections; $x = \frac{1}{6}\pi+2n\pi$, maximum points; $x = \frac{1}{6}\pi+(2n+1)\pi$, minimum points.

9. (i) Maximum values at $\frac{1}{2}\pi+(2n+1)\pi$, $\tan^{-1}1/\sqrt{2}+2n\pi$, $-\tan^{-1}1/\sqrt{2}+(2n+1)\pi$; (ii) minimum values at $\frac{1}{2}\pi+2n\pi$, $-\tan^{-1}1/\sqrt{2}+2n\pi$, $\tan^{-1}1/\sqrt{2}+(2n+1)\pi$.

INDEX